ENVIRONMENTAL HEALTH - PHYSICAL, CHEMICAL AND BIOLOGICAL FACTORS

GRASSLANDS: TYPES, BIODIVERSITY AND IMPACTS

ENVIRONMENTAL HEALTH - PHYSICAL, CHEMICAL AND BIOLOGICAL FACTORS

Additional books in this series can be found on Nova's website under the Series tab.

Additional E-books in this series can be found on Nova's website under the E-books tab.

ENVIRONMENTAL HEALTH - PHYSICAL, CHEMICAL AND BIOLOGICAL FACTORS

GRASSLANDS: TYPES, BIODIVERSITY AND IMPACTS

WEN-JUN ZHANG

EDITOR

Nova Science Publishers, Inc.
New York

For permission to use material from this book please contact us:
Telephone 631-231-7269; Fax 631-231-8175
Web Site: http://www.novapublishers.com

Library of Congress Cataloging-in-Publication Data

ISBN 978-1-61470-555-0
Grasslands : types, biodiversity and impacts / editor Wen-Jun Zhang.
p. cm.
Includes bibliographical references and index.
ISBN 978-1-61470-555-0 (hardcover : alk. paper) 1. Grasslands. 2. Grassland conservation. 3. Grassland ecology. 4. Biodiversity conservation. I. Zhang, Wen-Jun.
QH87.7.G738 2011
577.4--dc23

2011024405

Published by Nova Science Publishers, Inc. † New York

CONTENTS

PREFACE

Grasslands, being one of most important ecosystems on earth, provide not only forage, livestock and fur, but also biodiversity and other ecosystem service functionalities. How to prevent grasslands from overgrazing, disertification, and other types of over-exploitation is a pressing issue. In recent years, grasslands have been dramatically changing as the changes of global climate and aggravation of human activities. Protection and reasonable use of grasslands are thus attracting more attention worldwide. This book provides researchers with diverse aspects of the latest advances in grasslands research. This book also covers a varied range of topics, for example, global biodiversity review, biotic interactions in the Rhizosphere, evaluation of soil quality, insect assemblages, orchard grass, soil microbial communities, grassland biodiversity management and soil nematodes.

Chapter 1 - Due to human disturbances, global biodiversity is rapidly losing. Long-term biodiversity monitoring has been conducting in the past years. Assessment of global biodiversity is a necessity. In present study large amounts of surveyed data on global biodiversity were collected and analyzed. Global situation of biodiversity loss and conservation, especially the situation in China, was reviewed and discussed.It was found that the total number of estimated species on earth is approximately eight times of the number of species described. Rainforests harbored the most diverse species in the world. Great risk for species extinction exists in these areas. Human disturbances to species have largely exceeded the natural selection. Less distribution areas, habitat destruction, unregulated logging, pollution and human hunting have been pushing the extinction of large numbers of plant and animal species in the world. Environmental legislation, green GDP, and other environmental concerned policies must be formulated and implemented by every country in order to prevent species extinction.

In the past surveys some organisms like insects, fish, non-vascular plants, reptiles and amphibians have not yet been attracted enough attention due to their difficulties to be sampled. Moreover, biodiversity surveys were not accurate enough, especially that for vascular plants. Sampling countries or regions were not reasonably distributed. Some important countries, such as Madagascar, Zaire, etc., have not been surveyed. All of these problems should be solved in the future surveys.

Chapter 2 - Today the authors recognise the *rhizosphere* as a biologically active zone of the soil around plant roots that contains different living organisms such as soil-borne microbes including bacteria, actinomycetes, fungi, algae, protozoa, invertebrates (collembolans, nematodes, earthworms) in their abiotic environment. Constant supply of carbon compounds from plant roots fuels complex interactions among rhizosphere organisms, including those between microorganisms and plants, among microorganisms, between animals and microorganisms, between animals and plants, and among animals. The animal-microbial-plant interactions in the rhizosphere are very complex, where indirect effects potentially play a significant role in structuring soil communities and, consequently, there is a growing appreciation and a need to start including indirect effects when studying biotic interactions in the rhizosphere. Therefore, interest of this chapter lies in the definition of indirect biotic interactions in the rhizosphere; the studies of grasslands soil food webs and their consequences for plant growth and importance for rhizosphere processes. Interactions between organisms that involve physical contact such as in predation and parasitism are said to be *direct*. As a consequence of these interactions, other organisms or resources are affected indirectly. *Indirect* interactions include any mechanism of interaction between species that is mediated through a number of steps, where one species affects another one without direct contact. The types of indirect interaction that occur within the rhizosphere can be classified by considering the nature of the interacting organisms and grouping them accordingly as microbe-fauna, plant-microbe, microbe-microbe, fauna-plant etc. This scheme seems more useful than any other in the context of the rhizosphere as the precise nature of the interaction is often difficult to identify and can more conveniently be considered in terms of the organisms involved. The authors propose a simplistic model of interactions to illustrate carbon fluxes in multi-dimensional relationships between microbial populations, plants, soil fauna, organic matter and exudates. Methods that are applied to study direct and indirect effects in soil food webs are also highlighted in this chapter.

Chapter 3 - Approximately one quarter of the agricultural surface is covered by grasslands. In the last decades there has been a dramatic increase in the productivity of many grasslands. Nowadays, intensive and traditionally managed grasslands coexist in many regions. Management changes have caused alterations, mainly in the nitrogen and phosphorus cycles and in the composition of the botanic and microbial communities of grassland soils. However, the impact of management changes on soil quality has seldom been evaluated. On the other hand, soil biochemical and biological properties are considered as suitable and quickly responsive parameters to estimate soil quality. My work wants to summarize the state of art of the study of soil quality in grassland soils, comparing different managed grassland and utilizing soil biochemical and biological properties as a tool to evaluate soil quality. This review will put a strong emphasis in the temperate grasslands located in NW Spain.

Chapter 4 - Orchardgrass is a widespread perennial grass, which is well-adapted to dry conditions and is suitable for silvopastoral systems due to its shade tolerance. The main environmental (temperature, nitrogen, water and shade) and management (regrowth duration) factors that affect morphology, physiology, dry matter (DM) production and nutritive value (crude protein, organic matter digestibility and macro-nutrient concentrations) of orchardgrass (*Dactylis glomerata* L.) in temperate climate are reviewed. Regrowth duration is a management factor that can be modified through the frequency and severity of defoliation (e.g. infrequent cutting for hay or silage, rotational or continuous grazing). The emphasis is on open pasture and silvopastoral systems conditions. This is followed by a review of how DM production could be predicted from a canopy photosynthesis model based on the photosynthetic capacity of leaves, the light intercepted by leaf surfaces (dependent upon canopy architecture and leaf area index, (LAI)). The predictive capability physiologically based pasture models makes them powerful tools for pasture management or in assisting agronomists to improve practices in pastoral or silvopastoral systems.

Chapter 5 - Grasslands are complex terrestrial ecosystems where interactions among producers (green plants), consumers (grazing animals) and decomposers (soil microorganisms) determine its productivity and functions. Although a number of management activities are practiced in the grassland systems to maximize the efficiency of conversion of solar energy or plant mass into animal product, effects of managements on structure and functions of the below-ground communities has not been well studied. The present chapter discusses about the effects of management history on plant and soil microbial communities in grasslands by comparing two types of semi-natural grasslands (tall-type by infrequent grazing and short-type by frequent grazing) and the improved grassland in northern part of Japan. Plant species richness was highest in the tall-type grassland and lowest in the improved grassland. Short-type and tall-type grasslands were dominated by the proportions of C_4 and forbs species, respectively, while the improved grassland was by C_3 species. Microbial communities were studied by phospholipid fatty acid (PLFA) profiling and rDNA finger printing methods. Major soil microbial communities including total microbial PLFA, mycorrhizal PLFA, and the saprophytic fungal PLFA showed significant differences among the three grassland types. Fungal DNA band number also differed significantly among the grasslands. However, significant differences in community composition of microbial groups mostly appeared between semi-natural tall-type and the improved grasslands indicating the role of human management on structuring soil microbial communities. All these results also suggest that a complex interaction of human management and grazing might have altered the structure of plant and soil microbial communities in grasslands.

Chapter 6 - Grasslands are vital landscape elements in Europe. Recently, the 180 million hectares of grasslands have a crucial role in maintaining the landscape level biodiversity. Alkali grasslands are typical in Central- and Eastern part of Europe, with large areas in the Carpathian-basin. These types of grasslands were not the most favorable targets of arable farming, but large areas affected by mineral fertilization, drainage, soil melioration and/or commercial seeding in the last 60 years. In the authors paper the authors present important vegetation characteristics, species composition and management of five different grassland types from the open annual alkali pioneer swards to tall grasses dominated wet alkali meadows. In general, alkali grasslands are usually species poor communities characterized by short (*Festuca pseudovina*, *F. rupicola*, *Poa angustifolia*) or tall grasses (*Alopecurus pratensis, Elymus repens*). They harbor several steppe endemics (e.g. *Plantago schwarzenbergiana*, *Cirsium brachycephalum*, *Limonium gmelinii* ssp. *hungarica*, *Puccinellia limosa* and *P. peisonis*) and halophyte species (*Salicornia prostrata, Salsola soda, Suaeda pannonica, S. maritima,*), adapted to high salt contents of soil. According to the uneven pattern of soil salt and water alkali grasslands are spatially very diverse. Maintaining alkali grasslands the extensive grazing mostly by cattle and sheep is essential. Nowadays, in large areas of alkali grasslands former grazing are ceased or replaced with mowing. This resulted in a change of species composition, decreased richness and/or litter accumulation. Alkali grasslands are refugees of alkali steppe vegetation, thus, restoration and preservation of their biodiversity have a high conservation priority in Habitats Directive of the EU (Pannonic salt steppes and salt marshes, 1530).

Chapter 7 - Grasslands are the most widespread ecosystem worldwide. They are the natural habitats of a multitude of the grasses and herb species. They are of particular economic and ecological importance. Grasslands, as a part of natural ecosystem, represent the natural habitat for many soil microorganisms and animal species. An important part of the natural microfauna of grass ecosystems are nematodes, which are a complex of morphologically diverse species. According to their life strategy, the soil nematodes are divided into two groups: free living soil species and the plant parasitic species. Plant composition in grasslands plays a key role in determining soil nematode composition both in above and belowground resource-based mechanisms and in altering abiotic conditions. Nematodes can also play an essentials role in the retention of grasslands ecosystems. As primary consumers of saprophytic bacteria and fungi, nematodes make mineral nutrients available to higher plants. Parasitizing on grass roots, plant parasitic species affect the viability of grass plants, due to reduction of water and plant nutrient uptake, leading to nutrient deficiencies, while other nematodes are antagonists of organisms that negatively affect plant growth, and thus benefiting the plants. Nematodes are omnipresent, various, abundant, in a direct contact with soluble compounds in the soil water through their permeable cuticle, easily extracted and divided into trophic or ecological groups differing in their food source that have developed during their evolution. Representation of nematode trophic groups in soil, species diversity and abundance of genera or species of nematodes within their community may serve as an indicator of the environmental assessment of land ecosystem based on ecological and diversity indices.

In: Grasslands: Types, Biodiversity and Impacts
Editor: Wen-Jun Zhang

ISBN: 978-1-61470-555-0

Commentary

HOW DOES MOWING OF GRASSLAND ON SEA WALL FLOOD DEFENCES AFFECT INSECT ASSEMBLAGES IN EASTERN ENGLAND?

Tim Gardiner*
Environment Agency, Fisheries,
Recreation and Biodiversity (FRB),
Iceni House, Cobham Road, Ipswich, Suffolk, UK

ABSTRACT

Sea wall flood embankments protect a large area of coastal land in eastern England (particularly in the county of Essex) from tidal flooding. These important defences also provide grassland habitats for scarce insects in Essex, including several declining bumblebees (*Bombus* spp.). The grassland of many sea walls is mown once annually to maintain the structural integrity of the defences; however, cutting also exerts an influence over insect populations. Many sea walls are mown in midsummer (July and August) which can lead to high mortality of insects in the sward and also remove forage resources for bumblebees in particular. A review was undertaken of recent case studies investigating the response of insect populations to various sea wall mowing regimes in Essex in eastern England. This review highlighted the importance of rotational mowing regimes for limiting damage to populations of the rare moth *Gortyna borelii lunata*; it seems that cutting sea wall grassland in strips allows this insect to persist on flood defences. Other small-scale studies indicated that leaving a strip of unmown grassland on the folding (or berm) on the landward side of a sea wall is essential for promoting high abundance and species richness of bumblebees and butterflies. It is possible that conservation management for insects may conflict with the annual mowing that may be needed to maintain high floristic diversity. Leaving sections of sea walls unmown for insects, particularly on the folding, may lead to a decline in the floristic diversity of the sward due to the build up of litter and development of tussocky grassland. Implementing a system of rotational mowing which incorporates unmown grassland on the folding could be a key step toward more environmentally sustainable sea wall maintenance regimes in Essex. It is also clear that more research is needed into the effects of sea wall mowing on insect abundance and diversity, as there have been very few replicated studies in the east of England.

---

* E-mail: tim.gardiner@environment-agency.gov.uk

INTRODUCTION

The Essex coast is one of the longest in England at approximately 480 km, the majority of the terrestrial land being protected from tidal flooding by sea wall flood defences (c. 450 km of sea wall) (Jermyn 1974). Hard surfaced sea walls provide little value for biodiversity, but the majority of walls are earth banks dominated by coarse grasses such as *Arrhenatherum elatius*, *Dactylis glomerata* and *Elytrigia atherica*. Grasslands on sea walls are predominantly unimproved (they have not been agriculturally improved by ploughing, fertiliser input or herbicide applications), and as such provide an extremely important wildlife resource in the intensively managed Essex countryside (Gardiner 2009c). Sea walls also form a continuous network of grassland habitat allowing species to disperse along the banks (Gibson 2000). Historically, in Essex, sea walls were erected to enclose salt marsh and make it suitable for agricultural production. They were built of marsh clay obtained from borrow pits on the landward side, or from the salt marsh on the seaward frontage (Institute of Estuarine and Coastal Studies 1992). Due to their small size the early sea walls in the 1800s were constantly raised and the grass was kept short to prevent the formation of tussocks which led to the establishment of bare earth. In the 1800s vegetation was often allowed to colonise newly constructed sea walls naturally from surrounding habitats (Institute of Estuarine and Coastal Studies 1992), which may indicate why they have developed unimproved grassland swards of a high nature conservation value, currently containing large insect populations on some stretches of coastline (e.g. Dengie Peninsula). However, early intensive management regimes leading to the creation of short grassland swards may have discouraged breeding insects in some instances, particularly those of tall grassland. The ecological value of sea wall habitats is often overlooked as they are considered the primary cause of salt marsh erosion in Essex due to the prevention of the landward migration of intertidal habitats ('coastal squeeze'). To mitigate for the erosion of salt marsh due to rising sea levels, there is a need for managed realignment involving the breaching of sea walls to form new areas of intertidal habitat (Dixon *et al.* 1998). However, the newly emerging Essex and South Suffolk Shoreline Management Plan (SMP) highlights that many sections of sea wall are to be maintained in their current state in the long-term ('hold the line') to protect significant human habitations (e.g. Canvey Island) or economically important farmland. Therefore, sea wall grassland is likely to remain as a significant coastal wildlife habitat for the next 100 years across much of the county and there is a substantial opportunity to increase the biodiversity value of these flood defences.

Due to the presence of unimproved grassland, sea walls provide sites where rare insect species can flourish on the Essex coast. Many sea walls are within or adjacent to areas that have conservation designations such as Sites of Special Scientific Interest (SSSIs), Special Areas of Conservation (SACs), Special Protection Areas (SPAs) and Ramsar wetlands. Insects such as the rare moth *Gortyna borelii lunata*, and its larval foodplant, *Peucedanum officinale*, are well known for their populations on sea walls in Hamford Water (Ringwood *et al.* 2004; Gardiner and Ringwood 2010). Sea walls are also important for plants of unimproved grassland such as *Genista tinctoria* (Tarpey and Heath 1990) and declining Red Data Book (RDB) and UK Biodiversity Action Plan (UK BAP) bumblebees such as *Bombus sylvarum* (Benton 2000). *Bombus sylvarum* needs extensive flower-rich areas and suitable nesting sites of long tussocky grass to survive (Benton 2000). Bumblebees such as *B.*

sylvarum are also late nesting species. Suitable habitats are often found on sea walls in Essex, which may be important corridors along which bumblebees can disperse. The loss of flower-rich habitats since the 1940s due mainly to agricultural intensification led to a steep decline in numbers of *B. sylvarum* (Edwards and Williams 2004), and it was largely restricted to a few locations along the Thames Estuary east of London, south Essex and north Kent by the 1990s (Plant and Harvey 1997; Harvey 2000). The bumblebee was primarily a species of 'brownfield' ex-industrial sites that were neglected, the benefit of this dereliction being the formation of some of the highest quality invertebrate habitats in the UK due to the mosaic of bare ground, flower-rich uncut grassland, and scrub that has established (Harvey 2000). However, these 'brownfield' sites are under intense pressure to be developed in the East Thames corridor in particular (Benton 2006), with numerous disused chalk pits being destroyed to allow housing estates to be built. In some areas of south Essex, sea walls form the last remaining flower-rich habitats due to the loss of 'brownfield' sites (Benton pers.comm.).

Rare and endangered bumblebee species are likely to continue to decline unless flower-rich foraging habitats are protected and sympathetically managed (Dicks *et al.* 2010). Concern has been expressed by naturalists that sea wall mowing regimes have not been favourable for rare insects on the coast in the past, often entire sea walls have been mown to a short sward height (< 10 cm) in midsummer (July and August), which is likely to completely remove bumblebee forage resources in one event, as well as leading to high mortality of foraging bees (Benton 2000) and other insects such as grasshoppers (Gardiner and Hill 2006). However, in the last five years there has been a rapid northward expansion in the range of both *Bombus humilis* and *B. sylvarum* away from the East Thames corridor in the extreme south of the county (Benton and Dobson 2006, 2007), with both species being present on the north-east Essex coast at Mersea Island in 2010, a shift in range north-eastwards of approximately 30 km. This range expansion has been entirely dependent on sea wall grasslands (Benton pers.comm.), suggesting that current mowing regimes must at least allow some forage and nesting habitat to remain on a yearly basis to facilitate their northward spread.

It would seem that sea walls can form important corridors that bumblebees utilise effectively when there are extensive habitats such as coastal grazing marsh on the landward side of the flood defences. Benton (2000) suggests that *Bombus muscorum*, a declining bumblebee nationally, can be found on sea walls in Essex, but it is not known whether flood defences can support populations of this bee alone when the inland habitats are unfavourable (e.g. arable farmland). Nevertheless, it is clear that sea walls can harbour important populations of scarce bumblebees, even if they only utilise the unmown areas as foraging habitat. The Essex coast in eastern England is rich in Orthoptera with 18 species (including the allied insects Dermaptera and Dictyoptera) being recorded on a regular basis (Gardiner and Ringwood 2010).

Several Nationally Scarce species have been recorded, including *Ectobius panzeri*, *Forficula lesnei* and *Platycleis albopunctata*. There are two species, *Conocephalus dorsalis* and *Tettigonia viridissima*, which are almost entirely restricted to the Essex coast, mainly utilising sea wall grassland and scrub. A third species, *Metrioptera roeselii* (also still Nationally Scarce), was particularly noted for its large populations on the Essex coast in the early 1900s, which was its main stronghold in the UK (Wake 1997). The insect has since spread to inland habitats since the 1940s, with a main period of range expansion from the

1980s onwards in response to climate change, increased habitat availability on farmland, and the high occurrence of long-winged (macropterous) individuals (Gardiner 2009c). However, sea walls remain an important habitat for this range expanding species, particularly as they still hold large 'donor' populations for its continued inland colonisation (Wake 1997). The aforementioned orthopterans all require tall grassland throughout the summer to maintain large populations, therefore, the influence of sea wall mowing on their abundance is likely to be important. It is the aim of this chapter to outline how insects, particularly scarce and endangered species of conspicuous groups such as bumblebees (Hymenoptera: *Bombus* spp.), butterflies (Lepidoptera) and grasshoppers/bush-crickets (Orthoptera), may respond to the mechanised mowing of sea walls in Essex, reviewing the available data, and suggesting management techniques to preserve the widest range of species.

Sea Wall Habitats

A typical sea wall in the county can be divided into five sections for the purposes of grassland management: borrowdyke edge (usually unmown), folding (mown regularly, vehicular usage), sea wall slope on the landward side of the raised embankment, footpath on the crest (mown regularly and trampled by walkers), and sea wall slope on the seaward side (usually unmown) (Figure 1). Salt marsh is often adjacent to the foot of the sea wall on its seaward side. Concrete footings may be present to reinforce the base of the wall.

Mown vs. Unmown

Both mown and unmown sea walls have biodiversity benefits. Initial research into the conservation management of the rare moth *G. borelii lunata* found that cutting annually of sea walls at the end of August is detrimental to the abundance of the species (Ringwood 2004a), probably because mowing at this time of year leaves the grass too short for the ovipositing female moths to lay their eggs on. It is recommended that where sites must be mown in August this is performed on rotation, leaving much of the site uncut each year (Ringwood 2004b). This moth prefers to lay its eggs on long, coarse grass species such as *A. elatius*, *D. glomerata*, and *Elytrigia* spp. (Ringwood *et al.* 2002b), which occur in close proximity to the larval foodplant, *P. officinale* (Ringwood *et al.* 2002a). The response of *G. borelii lunata* to a rotational sea wall mowing regime was monitored in Hamford Water in north-east Essex. The protocol for mowing was agreed with Natural England due to the listing of the moth in Schedule 5 of the Wildlife and Countryside Act (1981). There was an increase in larval occurrence on the sections mown between 10-25 August on a three-year rotation, and a decrease on the unmown strip next to the borrowdyke (Figure 2).

Before the mowing regime was established in 2005, larval feeding signs were concentrated on *P. officinale* along the borrowdyke edge. Consequently, it appeared that the revised rotational management resulted in *G. borelii lunata* utilising the sea wall more uniformly, rather than being largely confined to areas that were not mown, such as the borrowdyke edge. In 2007, the percentage of *P. officinale* plants with signs of larval feeding was considerably higher than in 2005, indicating that the rotational management was beneficial to the abundance of the moth on the sea wall. Scrub encroachment where sections of sea wall are uncut appears to be a significant problem for *G. borelii lunata* because *Prunus spinosa* can quickly smother the larval foodplant, *P. officinale* (Ringwood 2008). A targeted programme of scrub clearance may be necessary to maintain stands of *P. officinale* under rotational mowing regimes.

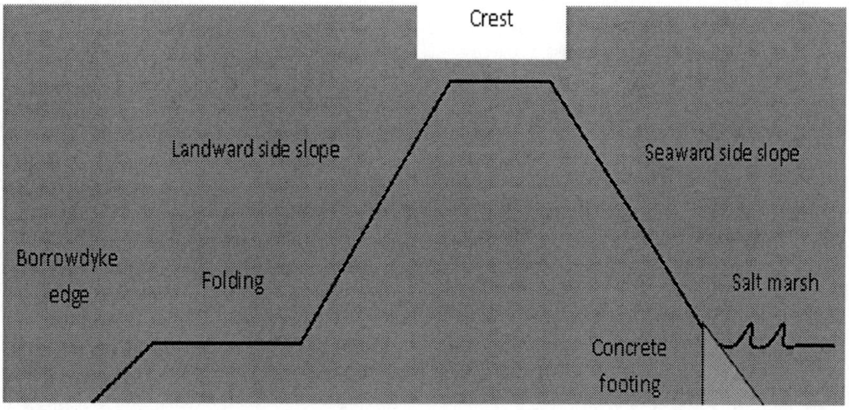

Figure 1. Cross-section of a typical Essex sea wall.

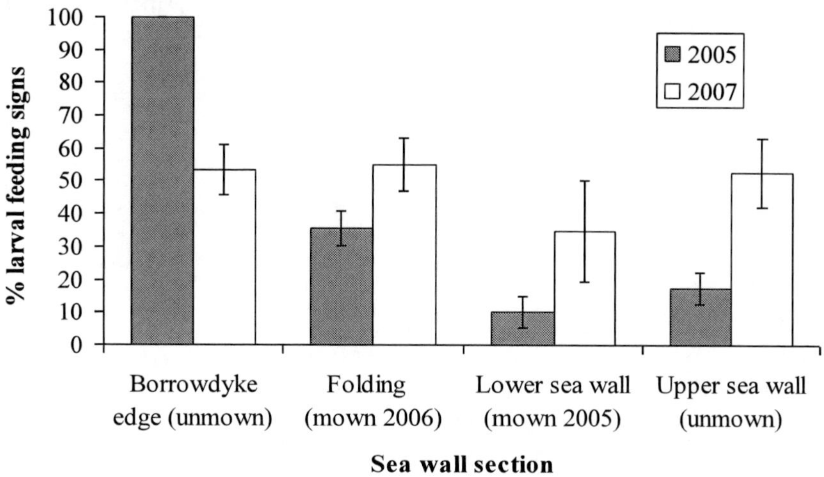

Figure 2. Mean percentage occurrence of larval feeding signs of *Gortyna borelii lunata* on *Peucedanum officinale* for four landward sections of a sea wall mown on a three-year rotation established in 2005 (standard error bars shown) (data: Ringwood 2008).

Along the Colne Estuary in north-east Essex (Lee-over-Sands), a standardised transect count (sampling method for bumblebee abundance followed Carvell *et al.* (2007), and for butterflies it adhered to Pollard and Yates (1993)) along a 1 km long section of mown and 1 km stretch of unmown grassland on a sea wall folding, and a corresponding 1 km transect along both mown and unmown sea wall crest, showed that the overall abundance and species richness of bumblebees and butterflies was higher on the section of folding that remained uncut during the spring and early summer (April-July) (Table 1). Indeed, where the folding had been mown in June, the abundance and species richness of bumblebees (1 spp.) and butterflies (6 spp.) was low, and the majority of insects were sighted along the crest of the wall which remained unmown at the time of the count (Table 1). Several scarce insect species were only present on the unmown folding, these included *B. muscorum* and *Melanargia galathea*. These observations suggest that mowing of the folding may be a key determinant of the abundance and species richness of indicator species such as bumblebees and butterflies. For the section of sea wall where both the folding and crest were uncut, it was clear that the folding had the highest abundance and species richness of both insect groups (Table 1), which indicated that abundant resources utilised by forage seeking insects persisted at the base of the wall in the absence of mowing. For bumblebees, there was an abundance of key summer forage plants such as *Lotus corniculatus* and *Trifolium pratense* on the uncut folding, which were absent from the section cut in June. It is also likely that insects displaced by the complete removal of their habitat from summer mowing of the folding may disperse to unmown areas of a sea wall, such as the crest in this instance, where there was a high species richness of butterflies (10 spp.) in particular.

Table 1. Total numbers of bumblebee and butterfly species counted along a transect at Lee-over-Sands on an unmown (control) sea wall (both folding and crest uncut), and an adjacent section of wall where the folding was mown in June but the crest remained uncut (1 km walked in all four sections, T. Gardiner unpublished data)

Order/species	Sea wall 1		Sea wall 2 (control)	
	Folding cut	Crest uncut	Folding uncut	Crest uncut
Bumblebees (Hymenoptera)				
Bombus lapidarius	2	17	4	6
*Bombus lucorum/terrestris**	0	12	1	2
*Bombus muscorum***	0	0	7	0
Bombus pascuorum	0	0	1	0
*Bombus ruderatus***	0	0	1	0
Bombus vestalis	0	8	2	2
Butterflies (Lepidoptera)				
*Coenonympha pamphilus***	0	1	4	0
Inachis io	0	0	1	0
Maniola jurtina	14	118	93	70
Melanargia galathea	0	0	1	0
Ochlodes sylvanus	2	70	78	1
Pieris brassicae	12	39	11	4
Pieris napi	0	3	1	1
Pieris rapae	0	4	3	0
Polyommatus icarus	0	2	0	0
Pyronia tithonus	9	223	40	55
*Thymelicus lineola/sylvestris**	11	132	98	17
Vanessa atalanta	1	1	0	0

*Species combined together when counting due to difficulty in distinguishing them apart in field surveys.

**UK Biodiversity Action Plan priority species.

At Paglesham along the Roach Estuary in south Essex, an experimental sea wall mowing regime was trailed in 2009 to ascertain whether leaving the folding uncut (apart from a mown 3 m wide strip to allow safe vehicle passage for cutting machinery) in summer during annual maintenance, was beneficial to bumblebees or butterflies. A standardised transect count (sampling method for bumblebee abundance followed Carvell *et al.* (2007), and for butterflies it adhered to Pollard and Yates (1993)) was undertaken in 2009 along unmown grassland on a sea wall folding and crest (1 km walked on both folding and crest), and a corresponding transect along an unmown folding and mown crest (1 km walked on both folding and crest) on an adjacent section of wall. The transect counts showed that the overall abundance and species richness of bumblebees (4 spp.) and butterflies (10 spp.) were higher on the folding than on the mown crest, where the former remained uncut in summer as part of the experimental cutting regime (sea wall 1; Table 2).

Table 2. Total numbers of bumblebee and butterfly species counted along a transect at Paglesham on an unmown (control) sea wall (both folding and crest uncut), and an adjacent section of wall where the folding was unmown but the crest was cut in June (1 km walked in all four sections, T. Gardiner unpublished data)

Order/species	Sea wall 1		Sea wall 2 (control)	
	Folding uncut	Crest cut	Folding uncut	Crest uncut
Bumblebees (Hymenoptera)				
*Bombus humilis***	2	0	0	0
Bombus lapidarius	31	6	0	5
Bombus pascuorum	2	0	0	0
*Bombus sylvarum***	7	0	0	0
Bombus vestalis	0	0	0	1
Butterflies (Lepidoptera)				
Aglais urticae	0	0	1	0
*Coenonympha pamphilus***	3	0	0	0
Inachis io	6	0	2	0
Lycaena phlaeas	0	1	0	0
Maniola jurtina	17	36	8	9
Melanargia galathea	0	0	1	1
Ochlodes sylvanus	212	3	17	24
Pieris brassicae	5	5	4	8
Pieris napi	1	0	0	0
Pieris rapae	0	0	0	2
Polyommatus icarus	1	0	0	0
Pyronia tithonus	46	98	22	51
*Thymelicus lineola/sylvestris**	128	7	12	5
Vanessa cardui	5	0	0	0

*Species combined together when counting due to difficulty in distinguishing them apart in field surveys.

**UK Biodiversity Action Plan priority species.

Indeed, on the control section of sea wall (sea wall 2), where the folding and crest remained uncut, there were much lower numbers of bumblebees and butterflies, and fewer species. The control sea wall has rarely been mown in recent decades and has developed a very tall and dense sward, with a large amount of litter and tussocky nature. This allowed very few forage plants for bumblebees and butterflies to persist in the sward (e.g. there was very little clover or *L. corniculatus*).

A study of bumblebees *Bombus* spp. (using transect method of Carvell *et al.* (2007)) on an inland flood embankment at Tilbury showed that a much higher number of bees was present on adjacent unmown banks compared to the mown flood wall (Table 3). Indeed, only three species were recorded from the mown flood embankment, compared to eight species on the unmown banks. Several *B. sylvarum* and *Bombus humilis* (both UK BAP priority species) were recorded foraging on the unmown embankments, despite the scarcity of forage resources such as *T. pratense* in the uncut grassland. Therefore, unmown grassland can be very important as habitat for bumblebees in certain circumstances, with tall herb species (e.g. *Cirsium arvense*) and scrub (*Rubus fruticosus*) being used by foraging bees. In the Tilbury study it was clear that the mown embankment had very little suitable habitat for bumblebees due to the complete removal of forage sources because of midsummer cutting.

The unmown embankments studied at Tilbury are also important for RDB plant species such as *Tordylium maximum*, a large number of plants (> 100) of which were discovered during the bumblebee surveys (Gardiner 2009a). It is likely that this population of *T. maximum* is the largest in the UK, with only two other sites currently known for this species in the wild in England. Therefore, the conservation of bumblebees by leaving unmown grassland on flood embankments can also promote the conservation of tall growing plant species in some instances. Benton (2000) suggests that unmanaged grassland which is not mown every year can form important nesting habitat for bumblebees, particularly scarce species such as *B. muscorum*. Bumblebee nests can also be destroyed by mowing in summer.

Table 3. Number of bumblebees *Bombus* spp. and *Tettigonia viridissima* recorded per km from standardised transect counts on a mown flood embankment and adjacent unmown banks in Tilbury Marshes in summer 2009 (T. Gardiner unpublished data)

Order/species	Bank management	
Bumblebees (Hymenoptera)	Mown	Unmown
Bombus hortorum	0.0	3.5
*Bombus humilis***	1.8	10.6
Bombus lapidarius	0.0	14.1
Bombus pascuorum	5.4	3.5
Bombus pratorum	0.0	2.6
*Bombus sylvarum***	0.0	5.3
*Bombus terrestris/lucorum**	0.0	75.0
Bombus vestalis	5.4	22.0
Bush-crickets (Orthoptera)		
Tettigonia viridissima	3.6	3.5

*Species combined together when counting due to difficulty in distinguishing them apart in field surveys.

**UK Biodiversity Action Plan priority species.

The locally scarce bush-cricket *T. viridissima* (an Essex Red Data List species; Gardiner and Harvey (2004)) was also heard stridulating along the unmown borrowdyke edge of both the mown and unmown embankments in similar abundance (Table 3). It was only present along the borrowdyke edge where there was uncut tall vegetation (this was the only unmown vegetation that remained on the cut wall) which it requires as habitat. Therefore, where there is an intensive summer mowing regime (two or three cuts), uncut grassland and scrub (difficult to cut on ditch edges) left along the borrowdyke banks should allow this scarce orthopteran to persist on sea walls.

Unmown grassland on a flood defence at Creeksea on the Crouch Estuary has the last remaining sea wall population of the localised beetle species *Lampyris noctiluca* in Essex. Larval surveys (using reptile survey mats) for this insect in July/August 2010 showed that isolated *Crataegus monogyna* bushes within open grassland on the sea wall are used by larvae, probably due to the presence of healthy snail populations upon which they prey (Table 4). Small patches of scrub (e.g. 1-2 m high bushes) form very useful edge (ecotone) habitat and the presence of them within unmown open grassland may be beneficial for the conservation of this declining beetle.

Table 4. Number of *Lampyris noctiluca* larvae, snails and reptiles recorded in different habitats on the unmown sea wall at Creeksea using reptile survey mats (felt roof tiles) in the summer of 2010 (data: Gardiner 2010)

	Open grass	Isolated scrub	Dense thicket
Lampyris noctiluca larvae	2	6	2
Snails	8	15	18
Reptiles	10	3	3

Management to establish a mosaic of habitats (e.g. grassland and scrub) may therefore be particularly important for *L. noctiluca* at Creeksea.

Despite the seeming preference of bumblebees, butterflies and *L. noctiluca* for grassland that is not mown in spring or summer, it appears that a lack of mowing on sea walls for many decades may lead to a loss of insect diversity associated with floristic resources. Even where the folding of a wall is to remain uncut, it is essential that occasional mowing takes place to preserve the abundance of forage resources such as clovers. It can be seen from quadrat surveys of a mown and unmown sea wall in Hamford Water that the floristic species richness was lower where cutting had not been undertaken (unmown spp. richness/m^2 = 4.9, mown spp. richness/m^2 = 5.9; Gardiner and Ringwood 2010). Key forage plants for bumblebees such as *T. pratense*, *Trifolium repens* and *Vicia cracca* were also absent from the unmown swards due to the tall and dense nature of the grassland on the sea walls, with an abundance of litter that smothered any low growing herb species (Gardiner and Ringwood 2010). The Essex Sea Wall Survey (Eco Surveys 1990) underlined the value of the folding for vascular plants likely to be used as forage sources by bumblebees (Table 5). Key forage species for *B. muscorum* such as *Lotus glaber*, *Ononis spinosa*, and *Trifolium squamosum* (Benton 2000), were predominantly recorded on the folding, with only a few observations of them from the landward/seaward slopes of the sea walls (Table 5). The folding is predominantly used for trafficking vehicles along a sea wall; therefore, it will receive occasional usage from tractor and off road vehicles, the wheels churning up the topsoil leading to the creation of bare earth and germination of plant species sensitive to soil disturbance (such as the Nationally Scarce annual plant species *Bupleurum tenuissimum*, *Hordeum marinum*, and *Puccinellia fasciculata*). Therefore, it is essential that some vehicles track along the folding each year (probably when mowing sea walls) to provide the soil disturbance required by these scarce plant species.

Table 5. Number of records of three plants used as forage by bumblebees and disturbance dependent species on differing sections of sea walls throughout Essex (source: Eco Surveys 1990)

Plant species	Borrowdyke edge	Folding	Landward/ seaward slope
Forage species for bumblebees			
Lotus glaber	15	47	13
Ononis spinosa	6	11	3
*Trifolium squamosum**	1	10	4
Disturbance dependent species			
*Bupleurum tenuissimum**	7	22	12
*Hordeum marinum**	0	68	24
*Puccinellia fasciculata**	9	26	0
Total number of records	38	184	56

* Nationally Scarce species.

A transect count of the number of patches (a patch is defined as a 20 x 20 cm area of a particular species) of four plant species on an unmown sea wall at Marsh House on the Dengie Peninsula revealed that all species were restricted to the folding in an absence of mowing (Table 6). Only one patch of *Galium verum* was present on the landward slope and all other species, including the preferred bumblebee forage plants, *L. glaber* and *T. pratense*, were only present on the folding (Table 6).

Table 6. Number of patches of four plant species (species counted within a 1 m band) from a transect count in August 2009 on the folding and landward slope (2 km walked on both sections) of an unmown sea wall at Marsh House on the Dengie Peninsula (T. Gardiner published data)

Plant species	Folding	Landward slope
Agrimonia eupatoria	68	0
Galium verum	52	1
Lotus glaber	19	0
Trifolium pratense	1	0
Total number of patches	140	1

This indicates that the apparent preference of declining bumblebees such as *B. muscorum* and *B. sylvarum* for the folding on sea walls on the Essex coast may be due to the greater number of forage plants on this section of the flood defences. The presence of plants such as *L. glaber* also indicates that there are abundant nectar resources for butterflies on the folding of sea walls.

Aspect

Recent work has emphasised that a species' habitat is made up of a number of discrete resources and that these resources may, or may not be, spatially separated (Ouin *et al.* 2004; Dennis *et al.* 2006; Dennis and Sparks 2006). Thermal microclimate resources (shelter and sunlight) have been found to be important to butterflies (Dover 1996; Corke 1997) and grasshoppers (Gardiner and Dover 2008) in the simplified landscape of arable farmland and may be important in other habitats such as sea walls. The orientation of sea wall embankments (e.g. north or south facing) and the shelter from the wind provided to the folding by the raised wall, could be crucially important in providing the sheltered microclimate that leads to 'sun trap' conditions utilised by Orthoptera in particular (Gardiner and Dover 2008).

The folding is likely to be sheltered from the prevailing south-westerly winds in Essex where sea walls run in an approximately east to west direction with the sea on the southern side of the flood defence (e.g. folding is on the 'leeward' north side of embankment and protected from prevailing south-westerly winds; Figure 3). The folding will be exposed to the prevailing wind direction on the Essex coast when a sea wall runs approximately north to south with the sea on the eastern side of the flood defence (Figure 3); therefore, the folding is on the 'windward' side of the embankment. The presence of scrub along the borrowdyke edge may provide shelter to insects on the folding where it is on the 'windward' less sheltered side of the embankment. It has been shown that species richness and abundance of Orthoptera are higher on the 'leeward' side of hedgerows and scrub (Gardiner and Dover 2008), so it is likely that the presence of linear strips of woody vegetation along the borrowdyke edge will provide the sheltered conditions needed for grasshoppers and bush-crickets on foldings which are exposed to the prevailing wind direction (e.g. those that run north to south; Figure 3).

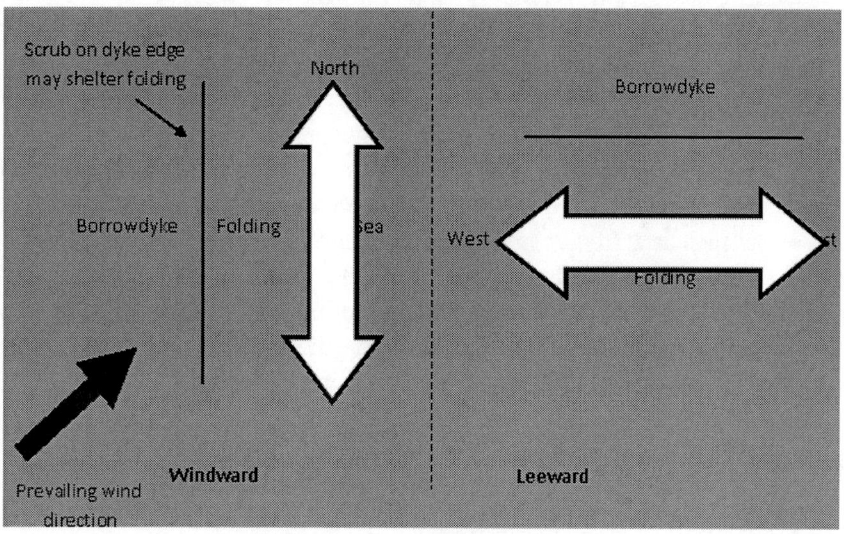

Figure 3. The possible impact of wind direction on coastal flood defences showing two scenarios: a 'windward' sea wall folding exposed to the prevailing south-westerly winds, and a 'leeward' folding sheltered from the wind by the raised embankment, the lack of shelter on the 'windward' folding may lead to lower insect abundance and species richness unless it is protected by scrub along the borrowdyke edge. The white arrows indicate the direction that the raised sea wall embankment runs.

Most orthopteran species in the UK do not fly strongly and only achieve flights over short distances of 2–3 m (Marshall and Haes 1988) but it seems that they are still likely to favour sheltered conditions similar to flying insects such as butterflies (Dover *et al.* 1997), particularly as the dispersal of some species such as *Chorthippus parallelus* (commonly found on sea walls) is reduced at high wind speeds (Gardiner 2009b).

However, east and south facing embankment slopes on sea walls (in the northern hemisphere) may also have more direct solar radiation in the late spring and summer (active insect season) in comparison to north and west facing slopes which will receive less direct sunlight.

This potentially places insects on the east and south facing slopes of sea walls at an advantage as they are able to begin feeding and moving earlier in the day than those on north and west facing slopes which receive less solar radiation until after 12:00 h. There are currently no data on the preferences of insects for sea wall slopes with differing aspects or foldings that are on the 'leeward' and 'windward' side of embankments. There is a need for further research on Essex sea walls to determine the likely importance of embankment orientation for the abundance and species richness of insects.

Grass Height and Flower Availability

Some insect species such as *L. noctiluca* (and spiders such as *Argiope bruennichi*) require tall, unmown grassland throughout the summer to maintain populations on sea walls in Essex. Scarce orthopterans such as *T. viridissima* require a complete absence of grass cutting and as such are likely to be restricted to the infrequently cut borrowdyke edge (as at Tilbury; Table 3) and seaward side of a wall, which receive little management or disturbance. Thick, undisturbed thatch (litter) is particularly valuable for invertebrates (Harvey and Smith 2006), but is probably undesirable from a flood defence aspect due to the difficulty in inspecting and maintaining the integrity of the sea wall.

Therefore, it is desirable that the borrowdyke edge and seaward side of sea walls (likely to provide tall grassland habitat) should remain unmown, unless it is absolutely necessary to clear scrub to maintain the integrity of the wall (tree roots can damage soil and lead to leakage of the bank). However, grass cutting may be essential to maintain floristic diversity and prevent plants such as *P. officinale* (larval foodplant of *G. borelii lunata*) from becoming smothered by scrub. Scarce bumblebees (e.g. *B. muscorum*) are likely to need flowering clovers (and other forage plants) throughout the spring and summer (March-September), whilst uncut nectar sources are important for butterflies (Dover *et al.* 2010). Therefore, leaving a section of sea wall uncut every year will increase the availability of forage for bumblebees (Harvey and Smith 2006) and nectar for butterflies.

Timing of Grass Cut

There is an increasingly large body of evidence which suggests that mechanised mowing of grassland from June-August (a traditional hay cut) has a devastating impact on insects that are present in the sward (Gardiner and Hill 2006; Gardiner and Hassall 2009; Dover *et al.*

2010; Humbert *et al.* 2010). For example, grasshoppers can be affected by cutting in hay meadows in two ways: through direct mortality caused by the cutting machinery (Wagner 2004; Gardiner and Hill 2006; Humbert *et al.* 2009, 2010) and by the creation of a thermally hostile environment with excessively high microclimate temperatures (often more than 44°C), which may necessitate the movement of grasshoppers to taller vegetation with a higher occurrence of shade habitat (e.g. tussocks), and lower sward temperatures which are nearer the 'optimum' temperature for growth and development (35–40°C) (Gardiner and Hassall 2009). Cutting may also remove all nectar sources for butterflies (Dover *et al.* 2010; Field *et al.* 2005; 2006; 2007) and forage resources (e.g. flowering clovers) for bumblebees (Benton 2000). Cutting from June-August can also lead to a high number of deaths in bumblebee populations (Benton 2000).

Research by Potts *et al.* (2009) revealed that bumblebees could be more abundant in habitats mown with a single July cut or no summer cut at all, than in more intensively mown plots (cut in May and July) which provides some evidence of the impacts of the timing of grassland mowing. However, in a replicated trial in the Netherlands (Kohler *et al.* 2007), an agri-environment scheme aimed at enhancing habitat for birds by delaying cutting had no impact on diversity or numbers of bumblebees *Bombus* spp. in wet meadow fields when compared with conventionally managed fields.

In Kohler's study, the absence of cutting during the bird breeding season (April-June) had no significant influence on bumblebee numbers or diversity, when compared with a treatment where mowing was not prohibited in the breeding season.

However, it is possible to alter the timing of grassland cutting so that the main period of insect activity is avoided; I would suggest cutting the sward in September to minimise the effect on invertebrates.

Despite the research by Kohler *et al.* (2007) showing that an absence of cutting in the bird breeding season does not lead to higher abundance of bumblebees, it is believed that their study may not necessarily be applicable to the situation on Essex sea walls where rarer species such as *B. humilis* and *B. sylvarum* are present (Kohler's study only considered the commoner *Bombus* spp.). Therefore, if cutting has to be undertaken from June-August, then a system of rotational mowing may be the best general option for insects, making sure to leave areas uncut every year as shelter for various species (particularly grasshoppers; Humbert *et al.* 2009, 2010). Generally, Essex sea walls are not cut in the bird breeding season (April-July inclusive) to avoid destroying bird nests (a criminal offence under the Wildlife and Countryside Act 1981) and disturbing waders with high tide roosts. It is suggested that this prohibition of mowing in the bird breeding season, on balance, probably aids the conservation of insects on Essex sea walls. Humbert *et al.* (2009, 2010) also suggests that the number of annual grassland cuts should be kept to a minimum (e.g. 1-2 cuts) to conserve Orthoptera.

Height of Grass Cut

Studies undertaken at Writtle College near Chelmsford, mid Essex (Gardiner and Hill 2005) indicate that grasshoppers spend most of their time in the ground zone (less than 20 cm from the soil surface), this suggests that they are particularly prone to being killed by mechanised cutting blades which cut hay between 5-10 cm from the soil surface. Gardiner and Hill (2006) confirmed that cutting for hay and silage with a rotary blade set at 9 cm in

July, does indeed lead to mortality of grasshoppers, including *Chorthippus albomarginatus*, which is commonly encountered on sea walls in Essex (Wake 1997). A cutting height of less than 10 cm should be avoided whilst mowing and is probably impractical on Essex sea walls due to the presence of ant hills and uneven ground. However, the study by Potts *et al.* (2009) found that pastures mown with a higher cutting height (10 cm instead of 5 cm) did not necessarily support more common *Bombus* spp. bumblebees than control plots managed for conventional silage production. It is believed that Pott's study (intensively managed farmland) probably does not relate to the situation on Essex sea walls as they only recorded commoner *Bombus* spp. and not the rare bees such as *B. humilis* and *B. sylvarum* which are likely to have nesting habitats destroyed by low cut heights on flood defences in the county.

Type of Mowing Machinery Used and Method of Harvesting

There have been very few studies on the effects of different types of mowing machinery on insects, and none specifically from Essex. However, we know that rotary mowers lead to mortality of grasshoppers through direct contact with the blades leading to decapitation or fatal damage to the abdomen of adults (Gardiner and Hill 2006; Gardiner 2009), particularly of *C. albomarginatus* and *C. parallelus*, two species commonly found on Essex sea walls. Indeed, studies have shown that mowing with a rotary blade can reduce *C. parallelus* adult density by approximately 63% in Essex hay meadows (Gardiner 2009b) and 57% in Switzerland (Humbert *et al.* 2010). The Essex study also suggested that the larger size of the mature *C. parallelus* adults made them particularly susceptible to being killed by rotary mowing blades, as cutting only reduced the density of the smaller sized *Chorthippus* spp. nymphs by 13% in a hay meadow (Gardiner 2009b). The impact of harvesting of cut material (generally through baling of hay) is likely to be as detrimental to Orthoptera as the actual mowing itself (Humbert *et al.* 2010), although it must be acknowledged that some live grasshoppers which are not killed by the cutting blades could be transported to other sites in the cut material (in green hay for example, Gardiner and Hill 2006). The bush-cricket *M. roeselii* (large populations on Essex sea walls) is negatively affected by hay mowing and subsequent raking/baling (mortality rate of between 74-91%) as is the Nationally Scarce *P. albopunctata* (88% mortality rate; Humbert *et al.* 2010), an insect which could potentially utilise favourably managed sea walls in the county (Gardiner *et al.* 2010). *Metrioptera roeselii* benefits from leaving grass cuttings on site ('in situ') to degrade naturally, as the cut vegetation provides cover for bush-crickets that survive mowing, and the eggs laid in grass stems would be removed by the process of hay making/baling (Gardiner 2009c).

Generally, Essex sea walls are cut using a combination of rotary and flail mowers. Flail mowers have the most detrimental impact on insect populations (e.g. greatest mortality) when compared with rotary cutting blades (Humbert *et al.* 2009). Interestingly, a front mounted flail (e.g. cutting equipment in front of tractor) is used to mow sea walls in eastern England, compared to the traditional rear mounted attachments (e.g. flail at rear of tractor), the former may give insects less time to move out of the path of the cutting flails leading to higher insect mortality, although this impact has not yet been proved. Whatever machinery is used to cut sea walls in Essex, if the mowing takes place during periods of peak insect abundance (e.g. June-August), then there is likely to be significant mortality of bumblebees and grasshoppers due to contact with the cutting blades/flails.

CONCLUSION: MANAGEMENT RECOMMENDATIONS FOR SEA WALLS IN ESSEX

Given the importance of sea walls for biodiversity in the county, recommendations for mowing management are likely to be highly beneficial to a wide range of protected and rare insect species. Table 7 summarises the likely effects of grass cutting and contains suggestions for improved mowing regimes.

Table 7. Likely impacts of various aspects of grass cutting on insects and suggestions for improved mowing regimes for Essex sea walls

Aspect of grass cutting	Likely effects	Improvements
Summer mowing (June-August)	High mortality of insects plus removal of forage/nectar sources and habitat, destruction of bumblebee nests	Mow late summer (September) or spring (March/April)
Autumn mowing (September-October)	Mortality of *G. borelii lunata* and removal of its egg-laying habitat	Where *G. borelii lunata* present mow between 10-25 August in 5 m wide strips
Low cutting height (blade set at < 10 cm)	High mortality of insects, destruction of bumblebee nests	Set cutting blade at a height greater than 10 cm
	Creation of uniform (even) and short sward height	Adjust cutting blade whilst mowing to create uneven sward height
Use of rotary and flail mowers	High mortality of insects, particularly grasshoppers and bush-crickets	Set cutting blade/flail at a height greater than 10 cm
Removal of cut grass after mowing (by baling)	Mortality of insects (particularly grasshoppers) during harvesting (baling), removal of eggs laid in grass stems	Leave cuttings 'in situ' as shelter for insects and to prevent the removal of eggs laid in grass stems
Mowing entire folding and banks of sea wall at one time from April-October	High mortality of insects plus removal of forage/nectar sources and habitat, destruction of bumblebee nests	Mow in strips, always leaving an unmown part of the sea wall every year
Not mowing sections of sea wall at all (e.g. borrowdyke edge, folding)	Development of litter/thatch and tussocks which are good for nesting bumblebees, spiders and *L. noctiluca*	Mow on long rotation (e.g. 2-4 years) to prevent loss of floristic diversity but still allow build up of litter/thatch

Having mown and unmown grass on a sea wall in the summer is likely to provide the tall grass refuges needed by Orthoptera, whilst maintaining the diversity of wildflowers needed as forage for bumblebees/butterflies and the integrity of the flood defence. Gardiner (2009d) suggested the implementation of a rotational mowing regime on some of the remoter sea walls in Essex.

Under this regime, the entire landward side of the sea wall would be cut over two years, instead of every summer. This cutting regime may be ideal for some Essex sea walls, particularly the remote ones, and will reduce the costs of mowing for the Environment Agency (the EA is a government body who maintain sea walls in Essex under their permissive powers). The mowing regime must also allow EA inspectors to survey the wall for damage at all times of year.

Currently, conservative estimates place the length of sea wall where mowing regimes take into account the presence of important insect assemblages at approximately 30-40 km (< 10% of sea wall on Essex coast). Many of the sea walls mown sensitively to benefit insect assemblages have unmown grassland on the folding as their main feature. On the east coast of the Dengie Peninsula, a continuous 8 km length of sea wall folding approximately 20 m wide has been left unmown every summer for several years, effectively establishing a 16 ha corridor that insects can disperse along. There is some evidence that leaving uncut grassland on sea walls on the Dengie has contributed to the north-eastwards range expansion of both *B. humilis* and *B. sylvarum* in Essex. The Dengie Peninsula is largely composed of intensively managed arable farmland, unlikely to be favourable for breeding bumblebees; therefore, the sea walls are the only favourable bee habitat over most of the Peninsula. Benton and Dobson (2007) reported probable breeding populations of *B. humilis* and *B. sylvarum* on the Dengie in 2005, in particular noting them from the east coast of the Peninsula where the 8 km stretch of folding had been left uncut (near Marsh House sea wall; Table 6). The sea wall here has *L. glaber* which is used as a foraging resource by *B. humilis* and *B. sylvarum* (Benton 2000). It is suggested by Benton and Dobson (2007) that suitable enhancement of habitat on the Dengie Peninsula may partially compensate for the loss of 'brownfield' sites in the East Thames corridor; therefore, sea walls provide the perfect opportunity to create 'bumblebee nature reserves' for the scarce and declining species such as *B. humilis* and *B. sylvarum*.

Monitoring of Insect Responses to Sea Wall Management: More Research Needed?

It is clear that in eastern England very little is known about how to manage sea walls to benefit the important insect populations they have, whilst maintaining their integrity as flood defences. It will be important to trial mowing regimes on sea walls in the east of England to ascertain how they work practically and to monitor the effects on key indicator species (such as rare bumblebees). This could take the form of simple and quick transect counts along 1 km sections of sea wall that are mown and those that are not. Such an approach would allow comparison of insect numbers between mowing regimes and years to allow any increases or declines in the biodiversity value of sea walls to be ascertained. As yet, we have no long-term data on the response of sea wall insects to different mowing regimes, this will be crucial in refining future management of these important habitats to benefit biodiversity. Any future studies should be fully replicated so that statistical analysis is possible.

ACKNOWLEDGMENTS

Thanks are due to the Fisheries, Recreation and Biodiversity (FRB), Operations Delivery and Asset Systems Management (ASM) teams of the Environment Agency for advice on sea walls and their mowing regimes. Thanks are also due to Professor Ted Benton of the University of Essex and Dr Zoë Ringwood of Natural England, who reviewed a draft of this chapter and made very useful comments. Zoë also kindly allowed the author to reproduce the *G. borelii lunata* data used in Figure 2.

REFERENCES

Benton T. and Dobson J. (2006) The Shrill Carder Bee on the move in Essex? *Essex Field Club Newsletter, 49,* 17-19.

Benton T. and Dobson J. (2007) Bumblebee report for 2006-7. *Essex Naturalist (New Series), 24,* 66-69.

Benton T. (2000) *The Bumblebees of Essex*. Lopinga Books: Wimbish.

Benton T. (2006) *Bumblebees: The natural history and identification of the species found in Britain. New Naturalist 98*. Harper Collins: London.

Carvell C., Meek W. R., Pywell R. F, Goulson D. and Nowakowski M. (2007) Comparing the efficacy of agri-environment schemes to enhance bumble bee abundance and diversity on arable field margins. *Journal of Applied Ecology, 44,* 29-40.

Corke D. (1997) *The Butterflies of Essex*. Lopinga Books: Wimbish.

Dennis R. L. H. and Sparks T. H. (2006) When is a habitat not a habitat? Dramatic resource use changes under differing weather conditions for the butterfly *Plebejus argus*. *Biological Conservation, 129,* 291-301.

Dennis R. L. H., Shreeve T. G. and van Dyck H. (2006) Habitats and resources: the need for a resource-based definition to conserve butterflies. *Biodiversity and Conservation, 15,* 1943-1966.

Dicks L. V, Showler D. A. and Sutherland W. J. (2010) *Bee Conservation: Evidence for the effects of interventions*. Pelagic Publishing: Exeter.

Dixon M., Leggett D. J. and Weight R. S. (1998) Habitat creation opportunities for landward coastal realignment: Essex case studies. *Journal of the Institution of Water and Environmental Management, 12,* 107-112.

Dover J. W. (1996) Factors affecting the distribution of satyrid butterflies on arable farmland. *Journal of Applied Ecology, 33,* 723-734.

Dover J. W., Rescia A., Fungarin˜o S., Fairburn J., Carey P., Lunt P., Dennis R. L. H. and Dover C. J. (2010) Can hay harvesting detrimentally affect adult butterfly abundance? *Journal of Insect Conservation, 14,* 413-418.

Dover J. W., Sparks T. H. and Greatorex-Davies J. N. (1997) The importance of shelter for butterflies in open landscapes. *Journal of Insect Conservation, 1,* 89-97.

Eco Surveys (1990) *Essex Sea Wall Survey*. Unpublished survey for National Rivers Authority: Basildon.

Edwards M. and Williams P. (2004) Where have all the bumblebees gone, and could they ever return? *British Wildlife, 15,* 305-312.

Field R. G., Gardiner T., Mason C. F. and Hill J. (2005) Agri-environment schemes and butterflies: the utilisation of 6 m grass margins. *Biodiversity and Conservation, 14,* 1969-1976.

Field R. G., Gardiner T., Mason C. F. and Hill J. (2006) Countryside Stewardship Scheme and butterflies: a study of plant and butterfly species richness. *Biodiversity and Conservation, 15,* 443-452.

Field R. G., Gardiner T., Mason C. F. and Hill J. (2007) Agri-environment schemes and butterflies: the utilisation of two metre arable field margins. *Biodiversity and Conservation, 16,* 465-474.

Gardiner T. (2009a) Hartwort *Tordylium maximum* discovered on Tilbury Marshes. *Essex Field Club Newsletter, 60,* 14.

Gardiner T. (2009b) *Hopping back to happiness? Conserving grasshoppers on farmland.* VDM Verlag: Saarbrücken.

Gardiner T. (2009c) Macropterism of Roesel's bushcricket *Metrioptera roeselii* in relation to climate change and landscape structure in eastern England. *Journal of Orthoptera Research, 18,* 95-102.

Gardiner T. (2009d) *Sea wall mowing for biodiversity.* Unpublished report for Environment Agency: Ipswich.

Gardiner T. (2010) *Glow-worm surveys and mitigation on a seawall at Creeksea.* Unpublished report for Environment Agency: Ipswich.

Gardiner T. and Dover J. (2008) Is microclimate important for Orthoptera in open landscapes? *Journal of Insect Conservation, 12,* 705-709.

Gardiner T. and Harvey P. (2004) Red Data List for Essex Orthoptera and Allied Insects. *Bulletin of the Amateur Entomologists' Society, 63,* 19-25.

Gardiner T. and Hassall M. (2009) Does microclimate affect grasshopper populations after cutting of hay in improved grassland? *Journal of Insect Conservation, 13,* 97-102.

Gardiner T. and Hill J. (2005) Behavioural observations of *Chorthippus parallelus* (Orthoptera: Acrididae) adults in managed grassland. *British Journal of Entomology and Natural History, 18,* 1-8.

Gardiner T. and Hill J. (2006) Mortality of Orthoptera caused by mechanised mowing of grassland. *British Journal of Entomology and Natural History, 19,* 38-40.

Gardiner T. and Ringwood Z. (2010) Species richness of orthopteroid insects and incidence of a rare moth on an island nature reserve threatened by sea level rise in the Walton Backwaters in eastern England. *Entomologist's Gazette, 61,* 251-261.

Gardiner T., Seago B., Benton T. and Dobson J. (2010) The use of bat detectors reveals a widespread population of Grey Bush-cricket *Platycleis albopunctata* at Colne Point and St Osyth naturists' beach. *Essex Naturalist (New Series), 27,* 209-213.

Gibson C. (2000) *The Essex Coast.... Beyond 2000.* English Nature: Colchester.

Harvey P. and Smith D. (2006) Howlands Marsh: a summary of terrestrial invertebrate survey in 2004. *Essex Naturalist (New Series), 23,* 77-88.

Harvey P. (2000) The East Thames Corridor: a nationally important invertebrate fauna under threat. *British Wildlife, 12,* 91-98.

Humbert J-Y., Ghazoul J. and Walter T. (2009) Meadow harvesting techniques and their impacts on field fauna. *Agriculture, Ecosystems and Environment, 130,* 1-8.

Humbert J-Y., Ghazoul J., Richner N. and Walter T. (2010) Hay harvesting causes high orthopteran mortality. *Agriculture, Ecosystems and Environment, 139,* 522-527.

Institute of Estuarine and Coastal Studies (1992) *Historical study of sites of natural sea wall failures in Essex*. Unpublished report by the Institute of Estuarine and Coastal Studies: University of Hull.

Jermyn S. T. (1974) *Flora of Essex*. Essex Naturalists' Trust: Colchester.

Kohler F., Verhulst J., Knop E., Herzog F. and Kleijn D. (2007) Indirect effects of grassland extensification schemes on pollinators in two contrasting European countries. *Biological Conservation, 135,* 302-307.

Marshall J. A. and Haes E. C. M. (1988) *Grasshoppers and Allied Insects of Great Britain and Ireland.* Harley Books: Colchester.

Ouin A., Aviron S., Dover J. and Burel F (2004) Complementation/supplementation of resources for butterflies in agricultural landscapes. *Agriculture, Ecosystems and Environment, 103,* 473-479.

Plant C. W and Harvey P. (1997) *Biodiversity Action Plan. Invertebrates of the South Essex Thames Terrace Gravels - Phase 1: Characterisation of the existing resource (3 volumes)*. Unpublished report number BS/055/96 for English Nature: Colchester.

Pollard E. and Yates T. (1993) *Monitoring Butterflies for Ecology and Conservation.* Chapman and Hall: London.

Potts S. G., Woodcock B. A., Roberts S. P. M., Tscheulin T., Pilgrim E. S., Brown V. K. and Tallowin J. R. (2009) Enhancing pollinator biodiversity in intensive grasslands. *Journal of Applied Ecology, 46,* 369-379.

Ringwood Z. (2004a) Fisher's Estuarine Moth: an Essex speciality. In: B. Goodey (Ed.) *The Moths of Essex*. Lopinga Books: Wimbish.

Ringwood Z. (2004b) The Ecology and Conservation of *Gortyna borelii lunata* (Lepidoptera: Noctuidae) in Britain. Unpublished PhD thesis, University of Essex: Colchester.

Ringwood Z. (2008) *The Application of Agri-environment Schemes to Secure the Future of Fisher's Estuarine Moth at a Landscape-scale, Final Project Report*. Unpublished report for Writtle College: Chelmsford.

Ringwood Z., Gardiner T., Steiner A. and Hill J. (2002a) Comparison of factors influencing the habitat characteristics of *Gortyna borelii* (Noctuidae) and its larval foodplant *Peucedanum officinale* in Britain and Germany. *Nota Lepidopterologica, 25,* 23-38.

Ringwood Z. K., Hill J. and Gibson C. (2002b) Observations on the ovipositing strategy of *Gortyna borelii* Pierret, 1837 (Lepidoptera, Noctuidae) in Britain. *Acta Zoologica Academiae Scientiarum Hungaricae, 48*, 89-99.

Ringwood Z., Hill J. and Gibson C. (2004) Conservation management of *Gortyna borelii lunata* (Lepidoptera: Noctuidae) in the United Kingdom. *Journal of Insect Conservation, 8,* 173–183.

Tarpey T. and Heath J. (1990) *Wild Flowers of North East Essex*. Colchester Natural History Society: Colchester.

Wagner C. (2004) Passive dispersal of *Metrioptera bicolor* (Phillipi 1830) (Orthopteroidea: Ensifera: Tettigoniidae) by transfer of hay. *Journal of Insect Conservation, 8,* 287–296.

Wake A. (1997) *Grasshoppers and Crickets (Orthoptera) of Essex*. Colchester Natural History Society: Colchester.

Paper reviewed by Professor Ted Benton of the University of Essex and Dr Zoe Ringwood of Natural England in December 2010.

In: Grasslands: Types, Biodiversity and Impacts
Editor: Wen-Jun Zhang

ISBN: 978-1-61470-555-0

Chapter 1

GLOBAL BIODIVERSITY LOSS AND CONSERVATION: A REVIEW

***WenJun Zhang**[*1,2] **and JianFeng Ou**[1]*

[1]School of Life Sciences, Sun Yat-sen University, Guangzhou, China

[2]International Academy of Ecology and Environmental Sciences, Hong Kong

ABSTRACT

Due to human disturbances, global biodiversity is rapidly losing. Long-term biodiversity monitoring has been conducting in the past years. Assessment of global biodiversity is a necessity. In present study large amounts of surveyed data on global biodiversity were collected and analyzed. Global situation of biodiversity loss and conservation, especially the situation in China, was reviewed and discussed.It was found that the total number of estimated species on earth is approximately eight times of the number of species described. Rainforests harbored the most diverse species in the world. Great risk for species extinction exists in these areas. Human disturbances to species have largely exceeded the natural selection. Less distribution areas, habitat destruction, unregulated logging, pollution and human hunting have been pushing the extinction of large numbers of plant and animal species in the world. Environmental legislation, green GDP, and other environmental concerned policies must be formulated and implemented by every country in order to prevent species extinction.

In the past surveys some organisms like insects, fish, non-vascular plants, reptiles and amphibians have not yet been attracted enough attention due to their difficulties to be sampled. Moreover, biodiversity surveys were not accurate enough, especially that for vascular plants. Sampling countries or regions were not reasonably distributed. Some important countries, such as Madagascar, Zaire, etc., have not been surveyed. All of these problems should be solved in the future surveys.

Keywords: biodiversity; loss and conservation; world; China; review.

[*] Corresponding author: zhwj@mail.sysu.edu.cn

1. INTRODUCTION

Biodiversity (biological diversity) means all types of species, genetic variants of species, and communities and ecosystems in which species interact with their environments (WCMC, 1994, 2000; Wang et al., 2010). In the Convention on Biological Diversity (1992), biodiversity was defined as the variability among living organisms from all sources including, inter alia, terrestrial, marine and other aquatic ecosystems and the ecological complexes of which they are part; this includes diversity within species, between species and of ecosystems. The contents of biodiversity grow with the advancement of biodiversity studies. For example, there are diverse variations for genes, cells, tissues, organs, populations, species, communities, ecosystems and landscapes. Up till now only genetic diversity, species diversity and ecosystem diversity have been studied in detail. Landscape diversity has also been attractions during the last decades, and some researchers suggested that landscape diversity should be treated as the fourth largest level of biodiversity (Ma, 1992). Some argued that biosphere diversity should also be considered a level in biodiversity hierarchy (Manuel and Molles, 1999).

Biodiversity is important and attractable for its values. The values of biodiversity are presently a hot topic. The values of biodiversity can be represented by direct use value, indirect uses value and potential uses value (Meng et al., 2004). Direct uses value means the bio-resources that can be directly consumed by human, which include foods, energy materials, medicinal resources and industrial raw materials, etc. Indirect uses value mainly refers to the functionalities concerning ecosystem functioning (Wilson, 1987; Ma and Qian, 1998; Chen and Ma, 2001; Zhang and Wei, 2009), including the functionalities for stabilizing ecosystems (Andow, 1991; Pimentel et al., 1992; Zhang and Barrion, 2006; Zhang, 2007), fixing solar energy, regulating hydrology and climate, absorbing or decomposing pollutants, storing nutrients and facilitating nutrient cycling, maintaining recreation, scientific research and public education, etc. Potential uses value, also called selection value, refers to the functionality for sustainably utilizing renewable resources, which are renewable if they have been rationally manipulated (Zhang and Yang, 2007).

Due to the effects of food webs, one plant species on earth will result in the extinction of 10-30 species of animals and microbes relying on this plant. Biodiversity loss will result in further deterioration of human environment and retard the social development, and even threat human survival (Ma and Qian, 1998; Li et al., 2009). In order to get a deep insight on global biodiversity loss and conservation, the present study collected and analyzed large amounts of surveyed data on biodiversity worldwide. The history and survey problems of biodiversity loss and conservation in the world and China were reviewed and discussed.

2. DATA SOURCES AND METHODS

Biodiversity data in this study were collected from biodiversity concerned online databases, journals and books. Most of the data were obtained from World Bank and the International Union for Conservation of Nature (IUCN). The biodiversity data of China were achieved from Chinese Academy of Sciences and Chinese Biodiversity Information System, etc. Raw data above were further summarized and synthesized to yield results and draw conclusions.

3. Analyses and Results

3.1 Profile of Global Biodiversity

How many species are on earth? This is a question plagued biologists for a long time. Now it is generally believed that earth harbors 14 million species, of which 1.75 million species have been scientifically described and named (Table 1). On the other hand, the exact number of scientifically described species and species considered to be effective are still unclear for most of taxonomic groups. The knowledge on the higher plant species and vertebrates has been clearly understood, but it is not true for insects, the lower invertebrates, fungi and some other groups. For example, until 1991 only 3058 bacterial species had been recorded and still huge numbers of bacterial species have not yet been recorded (WCMC, 1992).

Table 1. The number of world's major taxonomic groups (10000 species) (Heywood et al, 1995)

	Species	Estimated number of
Virus	0.4	40
Bacteria	0.4	100
Fungi	7.2	150
Protozoa	4.0	20
Algae	4.0	40
Higher plants	27.0	32
Nematodes	2.5	40
Crustaceans	4.0	15
Spders	7.5	75
Insects	95.0	800
Molluscs	7.0	20
Vertebrates	4.5	5
Others	11.5	25
Total	175.0	1362

Computed from Table 1, the estimated number of species on earth (N, 10000 species) is positively related to the number of species described (n, 10000species), which has the following relationship:

$$N=-0.52+7.80n$$
$$r2=0.9688,\ p<0.01$$

This means that the total number of estimated species on earth is approximately eight times of the number of species described.

3.2 Mass Species Extinctions in the History

It is well known that since the birth of earth there have been about five times of mass species extinctions, of which each extinction has almost led to the extinction of most species.

The first mass extinction occurred 440 million years ago, in the late Ordovician.

Because of the cooling of the global climate, earth was completely covered by glaciers. The coastal biosphere was destroyed, resulting in 85% of species extinction. Various invertebrates living in different water bodies disappeared during this period.

The second mass extinction occurred in the late Devonian, caused by cold climate and ocean retreat. The mass extinction led to the extinction of huge numbers of fish species.

The third mass extinction occurred 250 million years ago in the late Permian. The mass extinction was caused by some disaster. The disaster destroyed 95% of the marine species and nearly 70% of the terrestrial species (mainly some insects, pristine Reptilia and shark-like animals).

The fourth mass extinction occurred in the late Triassic. The period was a transition from the Paleozoic to the formation of modern biota. Earlier plants were mainly drought-tolerant species. As the climate changed to temperate and humid, more plant species appeared. Some plants similar to the modern time, e.g., pine, cycads, etc., also appeared. Meanwhile, those plants thrived in the Paleozoic disappeared gradually due to climate changes.

The fifth mass extinction occurred 65 million years ago. About 90% species on earth, including all dinosaurs, 20% of terrestrial vertebrates, 16% of marine species and almost all of the mammal species were destroyed during this period. However the causes of the extinction, especially sudden disappearance of dinosaurs, are still not clear.

Overall the causes of the above five mass species extinctions were abnormal climate changes or natural disasters. Apart from five major mass extinctions, a basic balance between species extinction and formation had been maintaining. During the past 200 million years, 90 vertebrate species would disappear every 100 years and one higher plant species would disappear every 27 years when human disturbances did not occur. However, since dinosaurs have been extinct especially human emerged on earth and started to disturb the natural environment, biodiversity loss on earth became to be faster than any time in the history. For example, the extinction rate of birds and mammals reached 100 to 1000 times of natural rate (Naeem et al., 1999).

According to a survey by IUCN, about 30% of the world's species would disappear in the next 50 years. Some researchers pointed out that if the extinction continues at the current rate, the sixth mass extinction would occur before the end of this century. By that time human survival will meet a great challenge.

3.3 Global Profile of Biodiversity Loss

According to a survey report for global flora and fauna, published by IUCN in 2009, among 47677 species of animals and plants investigated, 17291 species are threatened to extinct, of which 21% of mammals, 30% of amphibians, 12% birds, 28% of reptiles, 37% of freshwater fish, 35% invertebrates and 70% of plants are threatened to extinct (IUCN, 2009).

In << Red List of Threatened Species>>, there are 5490 species of mammals, of which 79% of species are endangered or extinct in the wild; 188 species are critically endangered;

449 species are endangered, and 505 species are vulnerable. In this red list, there are 1677 species of reptiles, of which 469 species are threatened to extinct; 22 species are endangered or extinct in the wild. In total 12151 species listed, 8500 species are threatened to extinct; 114 species are endangered or extinct in the wild (IUCN, 2009).

IUCN also noted that in the past five centuries, about 900 species of plants and animals had disappeared on earth. More than 10000 species are endangered to extinct now.

Twenty countries with the most endangered species, and the data on most endangered species are indicated in Table 2.

Table 2. Statistics of the most endangered species in top 20 countries

	Total	Mammals	Birds	Reptiles	Amphibians	Fish	Mollusks	Plants
Ecuador	2211	43	71	12	171	17	49	1848
United States	1203	36	73	32	56	212	273	521
Malaysia	1166	70	42	20	47	55	30	902
Indonesia	1126	184	154	26	33	135	4	590
Mexico	990	101	55	94	211	127	6	396
China	841	73	82	30	88	90	9	469
Australia	804	54	50	40	48	94	175	343
Brazil	769	82	134	21	30	80	20	402
India	687	75	75	24	66	61	2	384
Philippines	682	38	67	35	35	60	3	444
Colombia	658	52	90	15	15	38	0	448
Tanzania	607	33	41	5	5	145	19	359
Madagascar	581	63	34	20	20	80	24	340
Sri Lanka	546	30	13	10	10	40	0	443
Peru	545	52	94	6	6	15	0	372
Cameroon	534	41	14	3	53	50	4	369
Tailand	467	56	41	21	5	70	1	273
Papua New Guinea	448	41	35	10	11	40	2	309
Vietnam	417	53	37	27	17	45	0	238
South Africa	407	23	36	20	21	80	25	202

Synthesized from IUCN(2009).

From Table 2, we can find that the top 20 countries with the most endangered species lie in low latitude areas near the equator, and they distribute in the cluster way: in the Eastern Hemisphere it ranges from northern China to southern Australia, including 10 countries. In the Western Hemisphere it is from northern America to southern Brazil and six countries are included. The remaining four African countries sporadically distribute in Africa. Most of these countries are of tropical rainforest climate, tropical monsoon climate, tropical-subtropical humid monsoon climate. Because of the stronger solar radiation in tropical regions, earth surface receives much energy and a relatively high rainfall, which lead to a better condition for plant and animal growth. Therefore species are diverse and more endangered species distribute in these area.

Apart from the United States and Australia, the remaining 18 countries are developing countries. It can be found that Ecuador has the most endangered species (2211 species) in the world, seconded by the United States (1203 species) and Malaysia (1166 species). McNeely(1990) classified a few countries situated in tropical and subtropical areas, including Brazil, Colombia, Ecuador, Peru, Mexico, Zaire, Madagascar, Australia, China, India, Indonesia and Malaysia, as the megadiversity countries. These countries hold more than 60-70% of the world's biodiversity (Green Peace, 2004). Between-group correlation matrix, computed from Table 2, is as the following (***: $p<0.01$; **: $p<0.05$; *: $p<0.1$):

	Mammals	Birds	Reptiles	Amphibians	Fish	Mollusks	Plants
Mammals	1.00	0.68***	0.36	0.22	0.27	-0.18	0.05
Birds	0.68	1.00	0.10	0.07	0.20	0.02	0.17
reptiles	0.36	0.10	1.00	0.66***	0.41*	0.17	-0.11
Amphibians	0.22	0.07	0.66	1.00	0.12	0.10	0.54**
Fish	0.27	0.20	0.41	0.12	1.00	0.61***	-0.20
Mollescs	-0.18	0.02	0.17	0.10	0.61	1.00	0.11
Plants	0.05	0.17	-0.11	0.54	-0.20	0.11	1.00

It is obviously that there are significant correlations between mammals and birds, reptiles and amphibians, reptiles and fish, amphibians and plants, and fish and mollusks. The correlations have presumably been resulted from similar habitat requirements and food chain relationships, which reveals that habitat destruction is a major factor to endanger species. The top ten countries with the most species in major taxonomic groups are listed in Table 3. The ten countries will play a key role in global biodiversity conservation. By comparing Table 2 and Table 3, we can find that the majority of countries listed are the same. Combining the data of two tables, we can draw a profile of endangered species of major taxonomic groups in the major countries, as shown in Table 4.

3.3.1 Analysis of Globally Threatened Plants

As can be seen from Table 4, among mammals, birds, amphibians, reptiles and seed plants in major taxonomic groups, the endangered degree of seed plants is lower. It is lower than 3% and mostly between 1% and 2%. Ecuador, Malaysia and the United States, with the most endangered plant species as indicated in Table 2, have not been listed in Table 3 and thus absent in Table 4. Moreover, tropical rainforest countries, Ecuador, Malaysia and Indonesia, with the most endangered plant species (Table 2), have the highest endangered degrees in Table 4. This means that the most threatened areas for plant species are in tropical rainforest countries. Therefore, endangered degree of seed plants has been ignored considerably, which should be put a further focus in the future.

Table 3. Top ten countries with the most species in major taxonomic groups (McNeely et al., 1990)

Ranks	Mammals	Birds	Amphibian	Reptiles	Seed Plants
1	Indonesia	Colombia	Brazil	Mexico	Brazil
	515	1721	516	717	55000
2	Mexico	Peru	Colombia	Australia	Colombia
	499	1701	407	686	45000
3	Brazil	Brazil	Ecuador	Indonesia	China
	428	1622	359	600	27000
4	Zaire	Indonesia	Mexico	Brazil	Mexico
	409	1519	282	467	25000
5	China	Ecuador	Indonesia	India	Indonesia
	394	1477	270	453	23000
6	Peru	Venezuela	China	Colombia	South Africa
	361	1275	265	383	21000
7	Colombia	Bolivia	Peru	Ecuador	Indonesia
	359	1250	251	345	20000
8	India	India	Zaire	Peru	Venezuela
	350	1200	216	297	20000
9	Uganda	Malaysia	United States	Malaysia	Peru
	311	1200	205	294	20000
10	Tanzania	China	Australia/Venezuela	Tailand/Papua New Guinea	the former Soviet Union
	310	1195	197	282	20000

Since 1997, the World Bank has occasionally conducted biodiversity surveys in some countries. The surveyed species included mammals, birds and vascular plants. In total 36 countries or regions, of which 15 countries were from Asia, 3 from Africa, 3 from South America, 10 from Europe and 2 from Oceania, were surveyed. The records of vascular plant species were always stable, so the data of vascular plants in 2008 is listed, as in Table 5. It may be found that the extinction threat of vascular plant species is also serious in some countries, especially the countries with vast areas of rainforests.

Table 4. Endangered degrees of major taxonomic groups in major countries (%)

Countries	Mammals	Birds	Amphibians	Reptiles	Seed plants
Ecuador	--	4.81	47.63	3.48	--
United States	--	--	27.32	--	--
Malaysia	--	3.50	--	6.80	--
Indonesia	35.73	10.14	--	4.33	2.95
Mexico	20.24	--	74.82	13.11	1.58
China	18.53	6.86	33.21	--	1.74
Australia	--	--	24.37	5.83	1.49
Brazil	19.16	8.26	5.81	4.50	0.73
India	21.43	6.25	--	5.30	--
Colombia	14.48	5.23	--	3.92	1.00
Tanzania	10.65	--	3.69	--	--
Peru	14.40	5.53	--	2.02	1.86
Tailand	--	--	2.39	7.45	--
Papua New	--	--	--	3.55	--
South Africa	--	--	--	--	0.96

Tropical rainforest ecosystems have been degrading in recent decades. Rapid reduction of worldwide tropical rainforests is mainly caused by large scale of unregulated logging. The endangered plant species in Indonesia, for example, rank the third position in the world. Unregulated logging is serious in the nation. In the Province of West Kalimantan alone, 16 million hectares of tropical rainforest has disappeared due to unregulated logging. If this trend continues, the tropical rainforest in this province would completely disappear by 2040. The annual volume of illegal unregulated logging in Indonesia reached 50.7 million cubic meters, resulting in 3.42 billion US dollars of economics losses (Green Peace, 2004). In addition to unregulated logging, the other human activities such as unregulated logging for land reclamation, pollution and even urbanization process are threatening some endangered plant species. In recent years the large scale of disappearance of forests, especially rainforests with diverse species, has attracted worldwide attention. The majority of countries have begun to vigorously combat unregulated logging and to establish natural reserves and ecological parks in order to protect endangered plant species. Ecuador, the top-ranked country of endangered plant species, has established one of the world's most famous natural reserves in Galapagos Islands (Earth Report, www.tve.org). The action was proved to be a better way to avoid the conflict between environmental protection and local economic development. Currently the action to save rainforests is becoming a commercial behavior. For example, a US investment bank has cost 9 million US dollars to buy the tropical rainforests of Indonesia for making commercial deal.

Table 5. Number of endangered vascular plant species of some countries in 2008

	Countries	Total	Endangered		Countries	Total	Endangered
Asia	China	32200	446	North Amer.	Canada	3270	1
	China Hong Kong	--	6		Mexico	26071	261
	Banglades	5000	12		United States	19473	242
	India	18664	247	South Amer.	Argentina	9372	42
	Indonesia	29375	386		Brazil	56215	382
	Iran	8000	1		Venezuela	21073	68
	Israel	2317	--	Europe	Czech	1900	4
	Japan	5565	12		France	4630	7
	Myanmar	7000	38		Germany	2682	12
	Korea	2898	--		Italy	5599	19
	Pakistan	4950	2		Netherlands	1221	--
	Philippines	8931	213		Poland	2450	4
	Sri Lanka	3314	280		Russia	11400	7
	Tailand	11625	86		Spain	5050	49
	Vietnam	10500	146		Turkey	8650	3
Africa	Egypt	2076	2		UK	1623	13
	Nigeria	4715	171	Oceania	Australia	15638	55
	South Africa	23420	73		New Zealand	2382	21

Synthesized from <<World Development Indicators>>, World Bank, 2008.

3.3.2 Analysis of Globally Threatened Animals

It can be seen from Table 4 and 6 that the countries with the most mammals in the world generally have the higher endangered degree. Among eight countries to be studied, at least one-tenth mammals are endangered species. Moreover, one-third mammals in Indonesia are endangered species.

Among all of the animal species, the endangered degrees of bird species in eight countries surveyed, between 3% and 10%, are the smallest (Table 4 and 7), of which Indonesia has the highest endangered degree of 10.14%. However globally one-eighth of bird species are threatened to extinct and one bird species disappears every 10 years.

Amphibians are significant in four groups of animals investigated (Table 4).

From the smallest Tailand (2.39%) to the highest Mexico (74.82%), there are significant differences of endangered degrees between eight countries surveyed. Among these countries the endangered species of amphibians in Mexico (74.82%) and Ecuador (47.63%) are especially worry. One of two amphibian species sighted in Ecuador will be the endangered species. Further, three of four amphibian species sighted in Mexico are the endangered species. This is a horrible fact. The endangered degree of reptile species is relatively small in four animal groups.

It is between 3% and 13% for the eleven countries surveyed. Among these countries Mexico has a higher endangered degree of reptile species (13.11%).

Indonesia and Mexico are two countries with huge numbers of endangered animal species. From Table 2 and Table 3 we can find that the endangered mammal and bird species in Indonesia are the most abundant in the world. The country also holds the largest number of animal species and the fourth largest numbers of bird species in the world. Similarly, Mexico's number of endangered amphibian and reptile species are the greatest in the world. The country holds the largest number of reptile species and the fourth largest number of amphibian species in the world.

Most of Indonesia's mammals, birds and other animals live in the local rainforests. As mentioned above, Indonesia's rainforests disappear rapidly. As a consequence these animals are losing their habitats and threatened to extinct. Indonesia is situated at the junction of the two most active seismic zones, pan-Pacific seismic belt and the Eurasian seismic zone. Earthquakes occur frequently in this country, which push the extinction of animal species. Human disturbances, in particular unregulated logging, are the most important causes for the reduction of rainforests and the species extinction. For example, illegal coffee cultivation in Indonesia has destroyed rainforests lived by tigers, rhinoceros, elephants and other endangered animals, and has accelerated the process of extinction of these animals (Lin, 2007).

In addition to amphibians and reptiles, there is also large numbers of endangered fish species in Mexico. All of these animals at certain extent rely on water and water environments. Because of the transfer of pollution industry from US and the coarse pattern of economic development, environmental pollution in Mexico is serious. Moreover a high-pathogenic chytrid fungus, Batrachochytrium dendrobatidis, was found to strongly infect and kill amphibians in Mexico (Yu et al., 2006). The rapid spread and widespread outbreak of this disease is greatly threatening amphibians worldwide.

Table 6. Number of endangered mammal species in some countries

	Countries	2008		2006		2004		2002	
		Tot	Endangered	Tot.	Endangered	Tot	Endangered	Tot	Endangered
Asia	China	1801	351	502	80	394	79	394	76
	China HK	363	37	57	1	--	1	24	1
	Bangladesh	735	89	131	23	125	23	109	21
	India	1602	313	422	85	390	88	316	86
	Indonesia	2271	464	667	146	515	147	436	140
	Iran	656	75	158	21	140	22	140	23
	Israel	649	79	115	13	116	14	92	14
	Japan	763	190	171	37	188	37	132	37
	Myanmar	1335	118	288	39	300	39	251	36
	Korea	512	54	89	12	49	13	49	13
	Pakistan	820	78	195	17	188	19	151	18
	Philippines	812	253	222	50	153	50	153	50
	Sri Lanka	504	177	123	21	88	22	88	20
	Tailand	1271	157	300	36	265	37	265	34
	Vietnam	1116	152	279	41	213	40	213	37
Africa	Egypt	599	59	118	6	98	13	98	12
	Nigeria	1189	79	290	25	274	27	274	25
	South Africa	1149	323	320	29	247	42	247	41

Table 6. (Continued)

	Countries	2008		2006		2004		2002	
		Tot	Endangered	Tot.	Endangered	Tot	Endangered	Tot	Endangered
North Amer.	Canada	683	77	211	16	193	14	193	14
	Mexico	1570	579	544	72	491	70	450	69
	United States	1356	937	468	40	428	37	428	37
South Amer.	Argentina	1413	152	375	32	320	34	320	32
	Brazil	2290	343	578	74	394	81	394	79
	Venezuela	1745	166	353	26	323	26	305	25
Europe	Czech	474	39	88	6	81	8	--	8
	France	665	117	148	16	93	18	93	18
	Germany	613	59	126	9	76	11	76	12
	Italy	610	119	132	12	90	14	90	14
	Netherlands	539	26	95	9	55	10	55	11
	Poland	534	38	110	12	84	15	84	15
	Russia	941	153	296	43	269	45	269	42
	Spain	647	170	132	20	82	24	82	24
	Turkey	581	121	145	15	116	17	116	17
	UK	660	38	103	10	50	12	50	12
Oceania	Australia	1227	568	376	63	252	63	252	63
	New Zealand	424	124	73	8	--	8	--	--

Synthesized from <<World Development Indicators>>, World Bank.

Table 7. Number of endangered bird species in some countries

	Countries	2008		2004		2002	
		Tot	Endangered	Tot.	Endangered	Tot	Endangered
Asia	China	1221	82	1100	73	618	74
	China Hong Kong	306	20	76	11	--	11
	Bangladesh	604	23	295	23	166	23
	India	1180	79	923	70	458	72
	Indonesia	1604	121	1519	113	929	114
	Iran	498	18	323	13	293	13
	Israel	534	18	180	12	162	12
	Japan	592	53	250	34	210	34
	Myanmar	1047	41	867	35	310	35
	Korea	423	34	112	25	138	25
	Pakistan	625	30	375	17	237	17
	Philippines	590	70	395	67	404	67
	Sri Lanka	381	16	250	14	126	14
	Tailand	971	42	616	37	285	37
	Vietnam	837	41	535	35	262	37
Africa	Egypt	481	17	153	7	123	7
	Nigeria	899	9	681	9	286	9
	South Africa	829	36	596	28	304	28

Table 7. (Continued)

	Countries	2008		2004		2002	
		Tot	Endangered	Tot.	Endangered	Tot	Endangered
North Amer.	Canada	472	19	426	8	310	8
	Mexico	1026	57	769	39	440	39
	United States	888	71	650	55	508	55
South Amer.	Argentina	1038	55	897	39	362	39
	Brazil	1712	120	1492	113	686	114
	Venezuela	1392	25	1181	24	547	24
Europe	Czech	386	9	199	2	205	2
	France	517	15	269	5	283	5
	Germany	487	14	239	5	247	5
	Italy	478	15	234	5	250	5
	Netherlands	444	11	191	4	192	4
	Poland	424	12	227	4	233	4
	Russia	645	47	628	38	528	38
	Spain	515	20	278	7	281	7
	Turkey	436	14	302	11	278	11
	UK	557	10	230	2	229	2
Oceania	Australia	851	60	649	35	497	37
	New Zealand	351	74	--	--	--	63

Synthesized from <<World Development Indicators>>, World Bank

3.4 PROFILE OF GLOBAL NATURE RESERVES

There are many ways for the protection of biodiversity. The most conventional methods include in-situ conservation, ex-situ conservation and release into the wild (Meng et al., 2004; Tian, 2005). However, to establish nature reserves is considered the most direct and effective way. It can make full use of in-situ conservation, ex-situ conservation and release into the wild to conserve biodiversity (Figure 1). Profile of natural reserves in some countries and regions is indicated in Table 8.

Figure 1. Shennongjia National Reserve. With large numbers of plant and animal species, it is one of the most famous national reserve in China.

It can be found from Table 8 that Brazil, the United States, Russia and China have the largest areas of natural reserves. The establishment of nature reserves is not related to a country's level of economic development. For example, Bangladesh and Venezuela have similar levels of economic development. However Bangladesh's natural reserves are much less than Venezuela in terms of their total land area. The establishment of nature reserves is dependent upon economy, policy guidelines, technologies and environments of the country (Chen et al., 2004).

Both Table 2 and Table 8 include 11 countries, such as China, the United States and Indonesia. It is found from that the countries with many endangered species will also have the larger areas of natural reserves (Table 9). Besides Mexico, Brazil and Tailand, the area ratio of natural reserves of the remaining countries are positively related to their number of endangered species. This indicates that the countries with many endangered species generally pay more attention to the establishment of natural reserves.

Table 8. Natural reserves of some countries or regions in 2008

	Countries	Area (10000 km^2)	Resrved area/ Land area(%)		Countries	Area (10000 km^2)	Resrved area/ Land area(%)
Asia	China	110.1	11.8	North Amer.	Canada	62.9	6.9
	China HK	0.03	24.7		Mexico	9.9	5.1
	Bangladesh	0.1	0.5		U. S.	149	16.3
	India	15.6	5.3	South Amer.	Argentina	17.4	6.4
	Indonesia	26	14.3		Brazil	153.3	18.1
	Iran	10.6	6.5		Venezuela	64.4	73.1
	Israel	0.5	21.3	Europe	Czech	1.4	18.7
	Japan	5.2	14.3		France	1.6	3
	Myanmar	3.5	5.4		Germany	11.2	32
	Korea	0.4	3.6		Italy	3.2	11
	Pakistan	7.3	9.5		Netherlands	1	28
	Philippines	2.4	8.2		Poland	7	23.1
	Sri Lanka	1.8	27.3		Russia	128.7	7.9
	Tailand	8	15.7		Spain	4.6	9.3
	Vietnam	1.4	4.4		Turkey	2	2.6
Africa	Egypt	5.6	5.6		UK	6	25
	Nigeria	5.5	6	Oceania	Australia	74.5	9.7
	S. Africa	7.4	6.1		New Zealand	6.5	24.2

Synthesized from <<World Development Indicators>>, World Bank, 2008 .

Table 9. Area ratios and number of endangered species in some countries

Country	Number of endangered		Reserved area/Land area	
	Number	Rank	Area ratio (%)	Rank
United States	1203	1	16.3	2
Indonesia	1126	2	14.3	4
Mexico	990	3	5.1	10
China	841	4	11.8	5
Australia	804	5	9.7	6
Brazil	769	6	18.1	1
India	687	7	5.3	9
Philippines	682	8	8.2	7
Tailand	467	9	15.7	3
Vietnam	417	10	4.4	11
South Africa	407	11	6.1	8

Table 10. Endemic species of vertebrates in China (Biodiversity Clearing-House Mechanism of China, 2004)

	Known in China	Endemic species	Percent (%)
Mammalia	581	110	18.93
Aves	1244	98	7.88
Reptilia	376	25	6.65
Amphibia	284	30	10.56
Pisces	3862	404	10.46
Total	6347	667	10.5

3.5 China's Biodiversity

3.5.1 Biodiversity Profile

After Malaysia and Brazil, China's biodiversity ranks the third in the world. There are 30000 species of higher plants and 6347 species of vertebrates, accounting for 10% and 14% of the world respecively. There are also 599 types of terrestrial ecosystems (Biodiversity Clearing-House Mechanism of China, 2004). The major features of China's biodiversity include:

[1] Diverse species. China has the most abundant gymnosperms. It has 10 families and 250 species of gymnosperms. There are 6347 vertebrate species, accounting for 14% of the world.

[2] Many endemic genera and species. Endemic species of the higher plants are the most abundant (17300 species), accounting for 57% of the higher plant species in China.

There are 667 endemic species among 6347 vertebrate species, which accounts for 10.5% of the country's total (Table 10).

[3] Origin of flora and fauna was much ancient. Because most of China has risen to the land since the late Mesozoic, and it did not suffer glaciers of the Quaternary Ice Age, many areas of China to different degree maintained the ancient relics of the Cretaceous-Tertiary ages. For example, in existing 7 families of tsugae class in the world, China has now 6 families. Among animal species, giant panda, Chinese river dolphin, and Chinese alligator, etc., are all relics of the ancient relic species.

[4] Cultivated plants, domesticated animals and their wild relatives' germplasm resources are very abundant. China is the originating country of rice and soybean. There are 50000 rice and 20000 soybean varieties in China. China possesses more than 11000 species of medicinal plants, 4215 species of grass. The ornamental flower plant species with importance, native to China, are more than 30 genera and 2238 species. It has also the most abundant varieties of domesticated animals (1938 varieties and classes) in the world.

[5] Abundant ecosystem types. China owns various global terrestrial ecosystems such as forests, scrub, grassland and savanna, meadow, desert, alpine tundra and other types. Due to different climatic and soil conditions, they are further divided into 599 sub-types. The types of marine and freshwater ecosystems are also very complete.

Despite abundant biodiversity, China's situation of biodiversity loss is very serious. As indicated in previous tables, China has 841 endangered species, ranking the sixth in the world, of which include 73 species of mammals, 82 species of birds, 30 species of reptiles, 88 amphibians, 90 species of fish, 9 species of mollusks and 469 species of plants.

3.5.2 Endangered Plants in China

The authors summarized the factors threatening 3425 plant species in China and found some common or similar endangered factors as follows (Table 11 and 12):

[1] Habitat destruction: the habitats for plant growth have been degrading or destroying.

[2] Less distribution areas. If the distribution areas of vulnerable plant species are less than 10 then they are considered less areas distributed, and If the distribution areas of endangered plant species are less than 5 they are considered less areas distributed. Overall there are less distribution areas in China.

[3] Unregulated logging and collection. Unregulated logging for timber and firewood uses and excessive collection to plant and animal resources are still popular in China.

[4] Illegal trading for plant and animal resources occurs. For examples, some plants have been using for ornamental and medicinal values.

[5] Intra-specific characteristics of plant species, for example, slow growth rate, low seed production and low germination rate and poor expansion ability of endangered plant species etc., limit their recovery.

Table 11. Profile of plant species

	Species	Percentage (%)
Vulnerable (VU)	1686	49.23
Endangered (EN)	1062	31.01
Critically endangered (CR)	677	19.77

Table 12. Endangered factors for China's endangered plant species

Endangered factors	Vulnerable		Endangered		Critically endangered		Total	
	Species	Percent(%)	Species	Percent(%)	Species	Percent(%)	Species	Percent(%)
Habitat destruction	781	46.32	432	40.68	238	35.16	1451	42.36
Less distribution areas	1219	72.30	916	86.25	664	98.08	2799	81.72
Unregulated logging and collection	460	27.28	518	48.78	299	44.17	1277	37.28
Illegal trading	297	17.62	287	27.02	133	19.65	717	20.93
Intra-specific characteristics	146	8.66	117	11.02	88	13.00	351	10.25

Synthesized from Biodiversity Clearing-House Mechanism of China (2004)

Table 13. Endangered factors for endangered animal species in China

Endangered factors		Unrated		Level_I		Level-II		Total	
		Species	Percent(%)	Species	Percent(%)	Species	Percent(%)	Species	Percent(%)
Environ. factors	Habitat distru.	100	55.56	79	82.29	76	60.80	255	63.59
	Environ. Pollut.	50	27.78	16	16.67	21	16.80	87	21.70
Human hunting	For foods	61	33.89	27	28.13	33	26.40	121	30.17
	For medicines	23	12.78	14	14.58	29	23.20	66	16.46
	Pests	2	1.11	6	6.25	2	1.60	10	2.49
	Others	77	42.78	47	48.96	41	32.80	165	41.15
Intra-specific bio-characteristics		50	27.78	23	23.96	14	11.20	87	21.70

Synthesized from Endangered and Protected Species Database of Chinese Animals (2001)

As can be seen from Table 12, the main cause for endangering plant species is less distribution areas. It can be found that with the increase of endangered degree, this factor becomes more important. For critically endangered plant species this factor accounts for 98% of importance.

The endangered factor, unregulated logging and collection, is overall ranked the third. However for the endangered (EN) and critically endangered (CR), it ranks the second. This represents a general trend similar to that in most of the rainforests, i.e., unregulated logging is an important endangered factor for the disappearance of biodiversity in China.

Habitat destruction has been increasing with the rapid development of China's economy. For example, vast areas of natural forests have been deforesting for cultivation of Eucalyptus in order to provide the materials of paper industry.

About 20.93% of endangered plant species in China were used for trading. Most of them were used as medicinal materials, which greatly speeds up the extinction of endangered plant species.

The importance of intra-specific characteristics of plants increases with the rise of endangered degree. It demonstrated that natural selection is the major endangered factor for highly endangered plant species. For less endangered species, human disturbances are important endangered factors. Generally this factor is less important among all of the endangered factors.

The correlation analysis, based on Table 12, revealed that correlations between vulnerable, endangered, and critically endangered were significant (vulnerable-endangered: $r=0.92$, $p<0.01$; vulnerable-critically endangered: $r=0.98$, $p<0.01$; endangered-critically endangered: $r=0.99$, $p<0.01$). These results demonstrate that the species of vulnerable, endangered and critically endangered suffer the same influences from endangered factors and the endangered factors are stable in the transitions of vulnerable to endangered and endangered to critically endangered.

3.5.3 Endangered Animals in China

We analyzed the endangered factors for 640 endangered animal species recorded in Endangered and Protected Species Database of Chinese Animals (2001) based on the standard of 3 protection levels (unrated, level-I, level-II). In this database in total 401 species were given the endangered factors, including 180 species unrated, 96 species for level-I, 125 species for level-II. We summarize the common or similar endangered factors as follows (Table 13):

[1] Environmental factors, include habitat destruction and environmental pollution.
[2] Human hunting, include poaching for food, for pharmaceutical ingredients, as pests to be hunted, as well as other excessive hunting activities.
[3] Intra-specific biological characteristics.

Different from endangered plant species, the importance of intra-species characteristics of animal species accounts for 21.7% of the total, which is resulted from the higher trophic level of animals (Zheng, 2001). Moreover, the influence of intra-specific characteristics declines as the increase of endangered degree of animal species.

Figure 2. South China Tiger Panthera tigris amoyensis.

It is no doubtful that environmental deterioration and human hunting are two important endangered factors threatening animal species. As the rapid process of urbanization and industrialization, many natural environments in China have been transformed to human uses. Many animal species, such as giant panda, are threatening to extinct due to habitat destruction and some other reasons (Lin et al., 2005). In addition, environmental pollution also threats the survival of animals. Human hunting especially for food and medicinal uses has incurred the possible extinction of some animal species, such as a known tiger, Panthera tigris amoyensis (Fig. 2).

4. Discussion

From the results above we can find that rainforests harbor the most abundant species in the world. Potential threat of species extinction in these areas is also great. Generally human disturbances on species have considerably exceeded the natural selection. Less distribution areas, habitat destruction, unregulated logging, environmental pollution and human hunting are pushing the extinction of large numbers of plant and animal species worldwide.

Environmental legislation, green GDP, and other policies must be formulated and implemented by every country in order to retard the extinction rate of species.

Some problems were significant in the surveys of global biodiversity as follows:

[1] On the global scale, the surveys of mammals, birds and vascular plants were relatively perfect, because they were economically important and easily surveyed. However, some other organisms, such as insects, reptiles, amphibians, fish and non-vascular plants despite their ecological and economical importance (Wilson, 1987; Andow, 1991; Pimentel et al., 1992; Altieri, 1995; WCMC, 1996; McGavin, 2000; Chen and Ma, 2001; Jia et al., 2006; IAEES, 2009), have not yet been surveyed due to their difficulties to be sampled.

[2] Surveys were not accurate enough, especially the surveys on vascular plants. Sincethe publication of the data in 2000, data have not significantly updated until 2008. There were some errors in the data surveyed. For example, Iran was recorded to have only one species of vascular plant.

[3] Sampling areas were not reasonably distributed. For example, only 7 in 12 countries with the most remarkable biodiversity have been surveyed. Some important countries, such as Zaire, Madagascar, etc., were not surveyed in detail.

Up till now there have not perfect, public and accurate survey reports on global biodiversity. On the other hand, biodiversity surveys in some developed countries have been ideally implemented and a lot of biodiversity databases were developed in these countries. For most of the developing countries the surveys were incomplete or even absent. For this reason, international cooperation under the guide of IUCN and other international organizations is necessary in the future.

ACKNOWLEDGMENTS

The authors are grateful to the help of Mrs. Xue Zhang, Mr. XiangRong Lǜ i, and Mrs. LiYun Huang. WenJun Zhang and JianFeng Ou contributed the same to the paper. The study is supported by the International Academy of Ecology and Environmental Sciences (2010-001).

REFERENCES

Altieri MA. 1995, *Agroecology: the Science of Sustainable Agriculture*. Westview Press.

Andow DA. 1991. Vegetational diversity and arthropod population response. *Annual Review of Entomology*, 36: 561–586.

Biodiversity Clearing-House Mechanism of China. 2004. *http://www.biodiv. gov.cn/ dyxxz/200402/t20040226_88574.htm.*

Chen LZ, Ma KP. *Biodiversity Science: Principles and Practices.* Shanghai Science and Technology Press, Shanghai, China, 2001.

Chen YC, Chen L, Sun J. 2004. Technical problems in the establishment of natural reserves. *North Environment*, 29(6): 34-36.

Convention on Biological Diversity. 1992. *http://www.cbd.int/convention/ convention.shtml.*

Database for Endangered Plant Species. 2004. *http://bd.brim. ac.cn /copy_of _159_226 _89_44/consult/chenys/SearchForm.asp.*

Earth Report. TVE international. *http://www.tve.org/.*

Endangered and Protected Species Database of Chinese Animals. 2001. *http://monkey.ioz.ac.cn/division/datasearch/index5.html.*

Green Peace. 2004. Tracing the illegal logger of rainforests. *Cultural Relics*, 4: 38-47.

Heywood VH, et al. *Global Biodiversity Assessment.* United Nations Environment Programme. Cambridge University Press, Cambridge, 1995.

IAEES. 2009. *http://www.iaees.org/environdata/environ.asp.*

Jia CS, Chi DF, Hu YY. 2006. Effects of forest plant communities on forest insect communities. *Journal of Anhui Agricultural Science,* 34(9): 1871-1872, 1883.

Li AQ. 1995. *Inner Mongolia Environmental Protection*, 7(4): 39-43.

Li YM, Niu D, Zhang ZQ. 2009. Review of international scientific programmes and frontiers of biodiversity research. *Acta Ecologica Sinica*, 29(4): 2115-2123.

Lin L. 2007. *Knowledge is Power*, 7: 52-53.

Lin YH, Gu HJ, Lon TL, et al. 2005. Impact of Logging upon the Habitat of Giant Panda in Pingwu County, Sichuan, China. *Scientia Silvae Sinicae,* 41(1): 109-115.

Ma KP. 1993. A discussion on concepts of biodiversity. *Chinese Biodiversity*, 1(1): 20-22.

Ma KP, Qian YQ. 1998. Biodiversity conservation and its research progress. *Chinese Journal of Applied and Environmental Biology*, 4(1): 95-99.

McNeely JA et al. *Conserving the World's Biological Diversity*. The World Bank, WRI, IUCN, WWF, 1990.

Manuel C, Molles JR. *Ecology Concepts and Applications*. McGraw Hill Companies, USA, 1999.

WCMC. Priorities for Conserving Global Species Richness and Endemism. *WCMC Biodiversity Series* No. 3, 1994.

WCMC. The Diversity of the Seas: a regional approach. *WCMC Biodiversity Series* No. 4, 1996.

WCMC. *Global Biodiversity: Status of earth's Living Resources*. Chapman and Hall, London, UK, 1992.

WCMC. *Global Biodiversity: Earth's living resources in the 21st Century*. World Conservation Press, Cambridge, UK, 2000.

McGavin GC. *Insects, Spiders and Other Terrestrial Arthropods.* Dorling Kindersley Limited, London, UK, 2000.

Meng XX, Feng JC, Zhou YJ, et al. *Conservation Biology*. Press of Minzu University of China, Beijing, China, 2004.

Naeem S, Chapin FS, Costanza R, et al. Biodiversity and Ecosystem Functioning: Maintaining Natural Life Support Processes. *Ecological Society of America*, Washington, USA, 1999.

Pimentel D, Stachow U, Takacs DA, Brubaker HW. 1992. Conserving biological diversity in agricultural/forestry systems. *Bioscience*, 42(5): 354-362.

Sheng XH. 2009. Natural, man-made interference and ecological restoration. *World Environment*, 3: 30-33.

Tian XJ. *Biodiversity and Conservation Biology*. Chemical Industry Press, Beijing, China, 2005.

Wang XM, Qu JS, Li YM, et al. 2010. Analysis on situation and tendency of international biodiversity research. *Acta Ecologica Sinica*, 30(4): 1066-1073.

Wilson EO. 1987. The little things that run the world. *Conservation Biology*, 1: 344-346.

World Bank. *World Development Indicators*, 2002-2008.

Zhang WJ. 2007. Computer inference of network of ecological interactions from sampling data. *Environmental Monitoring and Assessment*, 124: 253-261.

Zhang WJ, Barrion AT. 2006. Function approximation and documentation of sampling data using artificial neural networks. *Environmental Monitoring and Assessment*, 122:185-201.

Zhang WJ, Wei W. 2009. Spatial succession modeling of biological communities: a multi-model approach. *Environmental Monitoring and Assessment*, 158: 213-230.

Zhang X, Yang J. 2007. Biodiversity: causes of loss and conservation strategies. *Anhui Agricultural Science Bulletin*, 13(10): 69-70.

Zheng Z. 2001. Stydy on plant and animal genecology. *Yunnan Environmental Science*, 20(2): 12-14, 56.

In: Grasslands: Types, Biodiversity and Impacts
Editor: Wen-Jun Zhang

ISBN: 978-1-61470-555-0

Chapter 2

INDIRECT BIOTIC INTERACTIONS IN THE RHIZOSPHERE OF GRASSLANDS

***Natalia Ladygina*[*] *and François Rineau*[*]**
[*] Department of Biology, Lund University,
Sölvegatan SE-Lund, Sweden

ABSTRACT

Today we recognise the *rhizosphere* as a biologically active zone of the soil around plant roots that contains different living organisms such as soil-borne microbes including bacteria, actinomycetes, fungi, algae, protozoa, invertebrates (collembolans, nematodes, earthworms) in their abiotic environment. Constant supply of carbon compounds from plant roots fuels complex interactions among rhizosphere organisms, including those between microorganisms and plants, among microorganisms, between animals and microorganisms, between animals and plants, and among animals. The animal-microbial-plant interactions in the rhizosphere are very complex, where indirect effects potentially play a significant role in structuring soil communities and, consequently, there is a growing appreciation and a need to start including indirect effects when studying biotic interactions in the rhizosphere. Therefore, interest of this chapter lies in the definition of indirect biotic interactions in the rhizosphere; the studies of grasslands soil food webs and their consequences for plant growth and importance for rhizosphere processes. Interactions between organisms that involve physical contact such as in predation and parasitism are said to be *direct*. As a consequence of these interactions, other organisms or resources are affected indirectly. *Indirect* interactions include any mechanism of interaction between species that is mediated through a number of steps, where one species affects another one without direct contact. The types of indirect interaction that occur within the rhizosphere can be classified by considering the nature of the interacting organisms and grouping them accordingly as microbe-fauna, plant-microbe, microbe-microbe, fauna-plant etc. This scheme seems more useful than any other in the context of the rhizosphere as the precise nature of the interaction is often difficult to identify and can more conveniently be considered in terms of the organisms involved. We propose a simplistic model of interactions to illustrate carbon fluxes in multi-dimensional relationships between microbial populations, plants, soil fauna, organic matter and exudates. Methods that are applied to study direct and indirect effects in soil food webs are also highlighted in this chapter.

Keywords: rhizosphere; indirect effects; grassland; soil organisms; study methods.

1. Introduction

Rhizosphere processes are driven by photosynthetically fixed carbon, which is either directly transferred to symbiotic organisms or released as root exudates stimulating growth of rhizosphere microbial communities that in turn regulate carbon flow via mineralization and immobilization of soil organic matter (Paterson et al., 1997). The constant supply of carbon compounds from plant roots fuels complex interactions among rhizosphere organisms including those between microorganisms and plants, among microorganisms, between animals and microorganisms, between animals and plants, and among animals. In complex system as rhizosphere, carbon sequestration is affected by multiple factors as root productivity, root turnover rates, exudation, mycorrhizal colonization, soil characteristics, plant community structure and microbial activity (Singh et al., 2004). Microbial activity is affected through altered plant inputs, where a major component of plant input is derived from litter fall and root turnover. Inputs also derive from rhizodeposition (loss of carbon compounds from active root systems) which may account for up to 40% of photoassimilates (Paterson et al., 1997).

We propose a simplistic model of interactions to illustrate carbon fluxes in multi-dimensional relationships between microbial populations (M), plants (P), soil fauna (F), organic matter (O) and exudates (E), see figure 1. The principal effects resulting from these interactions may be summarized as follows:

F → M:Soil fauna is known to be the important grazer of the microflora in grassland ecosystems, feeding selectively on certain soil microorganisms. This may stimulate growth of the less preferred species due to a reduced competition between species (Alphei et al., 1996; Maraun et al., 1998).

M → F: Microorganisms can indirectly positively affect fauna via transformation and release of active substances (e.g. NH_4 or homoserine lactone) (Nikolyuk, 1969; Clarholm, 1994) or negatively via direct predation on the fauna (Jaffee et al., 2007).

E → M, F: Exudates provide carbon to soil microorganisms, which in turn provide nitrogen and phosphorus to the plant via mineralization and immobilization of organic matter. Exudates have been shown to increase the mass and activity of soil microorganisms and fauna (Butler, 2003).

O → M, F, P: Soil organic matter is formed by microbial action on plant litter and dead animals (Barea et al., 2004).

M → P: A high diversity of microorganisms has been identified to have deleterious, beneficial or neutral effects on plants. Microbial populations can produce biologically active substances, which affect plants either positively – growth promotion or negatively – growth inhibition (Butler, 2003).

F → P: Soil biota influence plants either directly by feeding on roots and forming different relationships with their host plants (Clarholm, 1994; Zwart et al., 1994; Wardle et al., 2004) or indirectly via dispersing microorganisms and their propagules in soil (Thorpe et al., 1993; Whipps and Budge, 1993; Williams et al., 1998; Knox et al., 2003).

P → M, F: The soil food web can control the succesional development of plant communities both directly and indirectly, and these plant community changes can in turn influence soil biota (Bever, 2003; Wardle et al., 2004). Plants affect microbial populations both positively (Grayston et al., 1998; Yang and Crowley, 2000; Grayston et al., 2001) and negatively (Nehl et al., 1996).

The term "rhizosphere" was first used in 1904 by Lorenz Hiltner, soil bacteriologist and professor of agronomy at the Technical College of Munich (Campbell and Greaves, 1990) to describe the interaction between bacteria and legume roots. The term has since then been modified and redefined. Today we recognise the *rhizosphere* as a biologically active zone of the soil around plant roots that contains different living organisms such as soil-borne microbes including bacteria, actinomycetes, fungi, algae, protozoa, invertebrates (collembolans, nematodes, earthworms) in their abiotic environment (Kennedy, 1998).

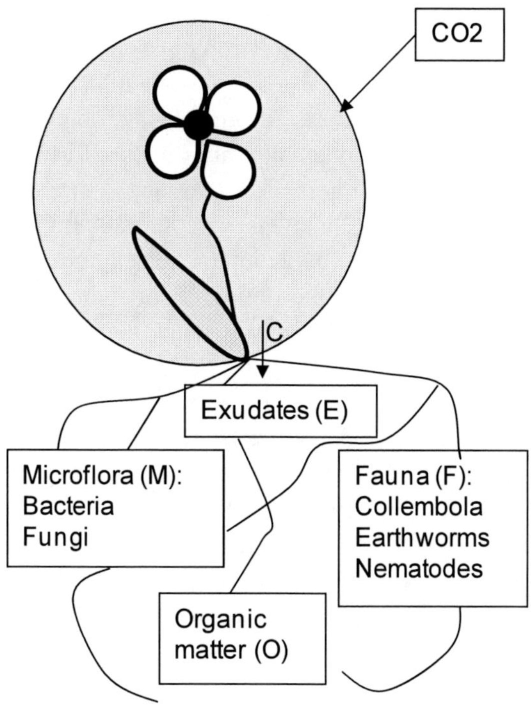

Figure 1. A simplistic model showing carbon flow in multi-dimensional interactions between aboveground plants and soil food web organisms.

The biotic interactions in the rhizosphere are very complex and variable, see figure 2. Rhizosphere processes are driven by large quantities of photosynthetically fixed carbon, which are either directly transferred to symbionts as mycorrhizal fungi, or released as root exudates stimulating rhizosphere bacterial growth (Werner, 1998; Toal et al., 2000). It has been stressed by soil microbiologists that beneficial and deleterious rhizosphere bacteria may significantly affect plant growth (Lynch, 1983). Due to the dual effect of root exudates, fueling a diverse bacterial community and intensifying the competition for nutrients between plants and microorganisms, the interaction between rhizosphere bacteria and plant roots can only be understood by also considering microfaunal predators (Clarholm, 1994). However, until now, exchange of thought and co-operation between the different disciplines investigating rhizosphere organisms is scarce. The view of plant physiologists is still focused on chemical exchange processes of nutrients in the rhizosphere (Loneragan, 1997). Similarly, soil microbiologists tend to neglect the existence of microfaunal-microbial interactions, or consider them only to be important in respect to plant pathogens (Gracia-Garza et al., 1997; Williams et al., 1998a,b). Soil zoologists very often overlook the diversity of rhizosphere organisms by focusing on faunal effects on microbial biomass (Bamforth et al., 1985; Ingham et al., 1986). Only recently the importance of micro-faunal grazing, not only for N-release from microbial biomass, but also for the composition of the bacterial community in the rhizosphere and the production of plant hormones, has been considered (Jentschke et al., 1995). Thus microbial grazers, especially protozoa seem to be more closely involved in rhizosphere interactions than previously assumed. Despite the crucial importance of interactions between roots (root exudates), microorganisms, invertebrates and plants, knowledge of these interactions is still fragmented and the mechanisms are poorly understood. In the past, plant physiologists, soil zoologists and soil microbiologists had been focusing on direct effects of microorganisms and soil animals in rhizosphere processes. However, due to the complex structure of animal-microbial-plant interactions in the rhizosphere, where indirect effects potentially play a significant role in structuring soil communities (Wardle et al., 2004), there is a growing appreciation of the community consequences of such indirect effects (Bukovinszky et al., 2008; Ohgushi, 2008), and a need to include indirect effects when studying biotic interactions in the rhizosphere.

The interest of this chapter lies in the definition of indirect biotic interactions in the rhizosphere; the studies of grassland soil food webs and their consequences for plant growth and importance for rhizosphere processes.

Interactions between organisms that involve physical contact, such as in predation and parasitism are said to be *direct*. As a consequence of these interactions, other organisms or resources are affected indirectly. *Indirect* interactions include any mechanism of interaction between species that is mediated through a number of steps, where one species affects another one without direct contact (Strauss, 1991). A scheme of a mechanism of an indirect interaction is presented in figure 3. In this interaction, an initiator species causes a change in a variable of a receiver species by changing one (or several) variables in one (or several) transmitter species. In contrast, no transmitter species is involved in a direct interaction (Strauss, 1991).

Indirect interactions are numerous, the following main categories have been distinguished and described since the 1960s (Menge, 1995): 1) keystone predation: a predator indirectly increases the abundance of competitors of its prey via consumption of the prey; 2) indirect mutualism: positively correlated changes in two species resulting from their respective predation on the competitor of the other's prey; 3) indirect commensalism: a potential indirect mutualist is more generalized in diet and also feeds on the main prey of the other mutualist; 4) habitat facilitation: an organism indirectly improves the habitat of a second one by altering the abundance of a third one; 5) indirect defense: the negative indirect effect of a nonprey species on a consumer species by harming the prey species of the consumer.

The types of indirect interaction that occur within the rhizosphere can also be classified by considering the nature of the interacting organisms and grouping them accordingly as microbe-fauna, plant-microbe, microbe-microbe, fauna-plant etc.

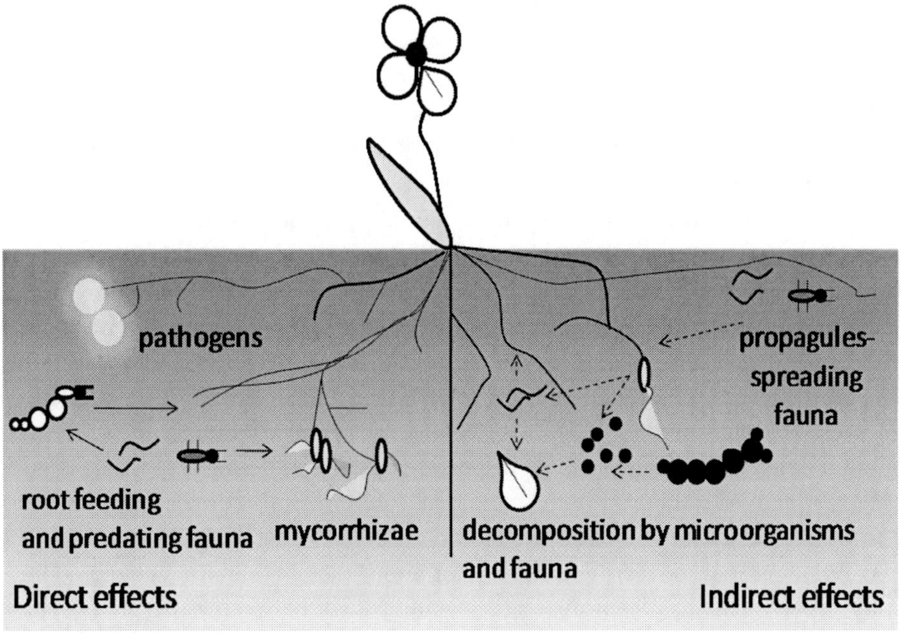

Figure 2. Rhizosphere food webs control plant community development and growth both directly (left) and indirectly (right), and plants in turn influence soil biota (right). Decomposition of litter and plant residues stimulate nutrient turnover, and therefore plant nutrient acquisition. Propagules-spreading fauna that disperse beneficial and pathogenic microorganisms both positively and negatively affecting plants (left). Soil organisms exert direct effects on plants by feeding on roots and forming beheficial and deleterious symbiotic relationships with their host plants. Direct effects are also in predation of soil organisms on each other.

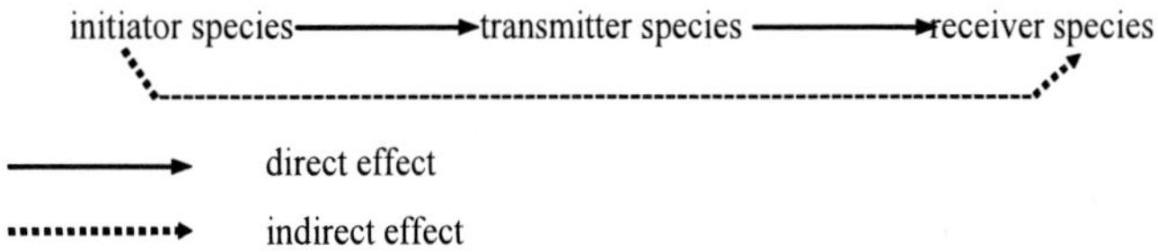

Figure 3. Scheme of a mechanism of an indirect interaction.

This latter scheme seems more useful in the context of the rhizosphere as the precise nature of the interaction is often difficult to identify and can more conveniently be considered in terms of the organisms involved.

2. Indirect Biotic Interactions in the Rhizosphere

2.1. Microbial–Faunal Indirect Interactions

Often soil fauna are studied in isolation from microorganisms and the significance of this is only just beginning to be appreciated. Soil fauna that operate at different spatial scales differ considerably in the nature of effects that they exert on the soil microorganisms.

Relationships between fauna and microorganisms are based on direct and indirect types of interactions. The feeding on microorganisms can be as an example of direct interactions. It is assumed that bacteria are often regulated by top-down effects of their grazers while fungi frequently are not (Wardle and Yeates, 1993). There are two reasons for this. First, fungi are morphologically relatively complex, contain a wide range of antifeeding (morphological and chemical) defenses, and have high carbon-to-nutrient ratios and therefore represent a poorer quality food source (Wardle and Yeates, 1993). And, secondly, fungal species differ tremendously in resource quality and fungal feeders are therefore less generalist in their feeding habits than are bacterial feeders (Newell, 1984).

Different faunal groups play different roles in microbial-faunal interactions: microarthropods interact with microorganisms as grazers and transporters, macroarthropods are primarly comminuters of litter, and saprophytic nematodes and protozoa are mainly bacteriovores (Williams et al., 1998). Invertebrates participate indirectly in soil processes by comminution of organic material and by dispersing bacteria and fungal spores (Schmidt et al., 1997; Madsen and Alexander, 1982; Knox et al., 2003). Soil animals are known to carry fungal propagules like spores, sporocarps or hyphae in their guts or on their cuticle (Thorpe et al., 1993; Williams et al., 1998). The carriage of propagules both in the gut and on the cuticle commonly occurs. This has been observed in interactions between *Aspergillus* spp. and the mite (*Acarus siro* L.), various fungi and *Coniothyrium minitans* and *Bradysia* sp. larvae, and a range of fungi with mixed species of collembola *Folsomia* sp. (Whipps and Budge, 1993; Williams et al., 1998). Collembola and oribatid mites are known to feed selectively on certain soil fungi. This may stimulate growth of the less preferred species due to a reduced competition between fungal species (Newell, 1984; Maraun et al., 1998). Sclerotia of *Sclerotinia sclerotiorium*, which had been grazed by the larvae of the fungus gnat *Bradysia coprophila* Lintner were degraded rapidly in soil infested with *Trichoderma hamatum* (Bon.) Bain (Gracia-Garza et al., 1997). Larval damage altered the sclerotia both physically and chemically, enhancing the activity of *T. hamatum*. In other studies, both collembolans and mites were shown to transmit conidia of the mycoparasite *Coniothyrium minitans* between sclerotia of *S. sclerotiorum* in soil (Williams et al., 1998b).

Soil fauna can interact with soil microorganisms via the "external rumen" relationship, which means that fauna transform litter in fecal pellets that are made up of much smaller particles than the ingested material. This transformed litter is highly favorable for microbial processes, particularly that of the bacterial component (Hamilton and Sillman, 1989).

Plant parasitic nematodes frequently facilitate the penetration of plant pathogenic fungi due to the damaging of cells by feeding activity. This synergistic effect may significantly intensify plant damage. Conversely, saprotrophic fungi have developed different strategies to exploit nematodes either by trapping or by endoparasitism and under certain conditions, this may indirectly reduce plant damage by parasitic nematodes (Jaffee, 1992). The feeding of roots by nematodes may increase allocation of carbon belowground and increase significantly the leakage of carbons from roots. Therefore, root-feeding nematodes may also stimulate rhizosphere microbial processes (Bardgett, 1998). Also, infection by arbuscular mycorrhizal fungi was found to increase the plant tolerance to nematode endoparasites (Tylka et al., 1991).

Another aspect of the indirect microbial-faunal interactions is the interaction between soil bacteria and protozoa. Bacteria can play important role in protozoan-plant interactions, mediated by hormones (e.g. homoserine lactone). Since many soil bacteria release plant hormones as a result of gaining metabolic energy from the transformation of hormonal precursors (e.g. tryptophan), it is possible that hormonal precursors are released by protozoa and are transformed into hormonal active substances by microorganisms (Nikolyuk, 1969).

2.2. Plant–Microbial Indirect Interactions

Studies on the interactions between plants and soil microorganisms in the rhizosphere are important for understanding a range of intrinsic processes, such as nutrient cycling, ecosystem functioning and carbon sequestration. On the base of current knowledge, indirect interactions between plants and microbes occur in the rhizosphere due to the loss of carbon by exudation, secretion or autolysis of older root cells of the plants. Root exudates have been shown to increase the biomass and activity of soil microbes and fauna found in the rhizosphere (Butler, 2003). Root exudates have a profound qualitative and quantitative effect on the rhizosphere microflora. The chemical components released by roots are classified into groups, depending on their mode of excretion: (1) water soluble exudates, such as sugars, amino acids, organic acids, hormones and vitamins, which leak from the root without the involvement of metabolic energy (Lynch and Whipps, 1990), (2) secreted substances, such as polymeric carbohydrates and enzymes which depend upon metabolic processes for their release (Gramss and Rudeschko, 1998), (3) lysates, released when cells autolyse, including cell walls, and, with time, whole roots (Curl and Truelove, 1986), (4) gases such as ethylene, CO_2 and CH_4 (Mitra et al., 2005), (5) plant mucilage (an exopolysaccharide/glycoprotein nature substance aiding in, for example, water storage) (Nagahashi and Douds, 2004), and (6) mucigel (a polysaccharide nature substance covering the rootcap of the plant roots) (Curl and Truelove, 1986). Roots also can liberate different antimicrobial peptides that act against various pathogenic fungi and oomycete (Odintsova et al., 2008; Van der Weerden et al., 2008).

Microbial populations can produce biologically active substances, which affect plant growth. These may be positive when these substances promote growth, as seems to be the case when biotin and pantothenic acid are produced or when iron uptake by the root is enchanced by agrobactin production by *Agrobacterium tumefaciens* (Lynch, 1990). The presence of homoserine lactone (HSL; a degradation product of N-acyl-homoserine lactone, a regulatory signal molecule that is commonly used by plant-associated bacteria) increases stomatal conductance and transpiration by 20-30% that help plants and rhizospheric bacteria to obtain diffusion-limited mineral nutrients, such as phosphorus (Bais, 2004). It is known that *Bacillus* species are prevalent in the rhizosphere, where they can produce plant growth factors and synthesize antifungal agents (Wipat and Harwood, 1999). Plant roots support the growth and activities of a wide variety of microorganisms that may have a profound effect on the growth and/or health of plants. Among these microorganisms, a high diversity of bacteria has been identified and categorized as deleterious, beneficial or neutral with respect to the plant. The beneficial bacteria, termed plant growth-promoting rhizobacteria (PGPR), are widely studied by microbiologists and agronomists because of their potential in plant production (Barea et al., 2004; Zahir et al., 2004). PGPR increase plant growth indirectly either by the suppression of well-known diseases caused by major pathogens or by reducing the deleterious effects of minor pathogens (microorganisms, which reduce plant growth but without obvious symptoms). PGPR may increase plant growth regulating ethylene production in roots, releasing phytohormones (Lynch, 1990) and decreasing heavy metal toxicity. For example, *Azospirillum*, a genus of versatile PGPR, is able to enhance the plant growth and yield of a wide range of economically important crops in different soils. Plant beneficial effects of *Azospirillum* have mainly attributed to the production of phytohormones, nitrate reduction, and nitrogen fixation (Somers, 2004). Mn (II) and Fe (II) are solubilized during the microbial production of acetic acid at low redox potentials, and thus are made potentially available to plants (Lynch, 1990). It is important to remember that deleterious rhizobacteria also can inhibit plant growth (Nehl et al., 1996). Some species of *Pseudomonas* produce cyanide, resulting in an indirect negative interaction with respect to the root. The microbial phenolic compounds (caffeic acids, myricetin etc.) could inhibit nitrification by *Nitrosomonas* and *Nitrobacter* in extremely low concentrations, potentially changing the form of availability of plant nitrogen (Lynch, 1990).

Additional interactions in the rhizosphere perform mechanism of control of the microbial community by the plants. A lot of the matter lost by the roots has low molecular weights and is easily decomposable by the microorganisms. This results in a changed selective pressure on the microbial population around the root (Curl and Truelove, 1986). But in general rhizosphere microorganisms are ecophysiologically adapted to these chemical conditions.

It is not conclusively known if plants actively promote beneficial soil microbial communities in their rhizosphere through rhizodeposition. Initial findings suggest that plants do promote specific bacterial groups in the rhizosphere regions (Grayston et al., 1998; Grayston et al., 2001). Most of the studies in which plants were grown in pots as a monoculture revealed that specific groups of microorganisms were associated with the rhizosphere (Baudoin et al., 2003). However, results from field studies with a mixed plant community found contradictory results (White et al., 2005). Some studies did not find such selection in the rhizosphere but other studies reported a strong correlation between plant and soil microbial community (Duineveld et al., 2001; Smalla et al., 2001). At the functional level, microorganisms that colonize the rhizosphere help plants to aquire phosphorus and potassium, and some enchance nitrogen uptake from the soil by their effect on root morphology and physiology (Cocking, 2003; Trolove et al., 2003). Recent observations of a higher diversity of ammonia-oxidizing bacteria (Briones et al., 2003) and nitrogen-fixing (*nifH*) genes (Cocking, 2003) in the rhizosphere when compared with bulk suggest that plants might promote functional groups rather then taxonomic groups of microorganisms. There is accumulating evidence that biotic interactions, occuring in the rhizosphere, play an important role in determining plant diversity aboveground by indirect effects on competing plants (Bever, 2003).

2.3. Faunal–Plant Indirect Interactions

The interaction between plant roots, root exudates and microorganisms can only be understood in relation to soil faunal activity, indicating that the soil fauna has an important function in regulating rhizosphere microbial processes and therefore significantly affect plant growth.

Microfaunal grazing may affect plant growth via nutrient effects by releasing NH_4 from consumed rhizosphere bacteria (Clarholm, 1994). Nutrients in the rhizosphere are taken up rapidly by soil bacteria and become only temporarily locked in bacterial biomass near the root surface and are successively liberated by microfaunal grazing (Bonkowski et al., 2000). The assumed mechanism, known as the "microbial loop" in soil (Clarholm, 1994), is triggered by the release of root exudates from plants that increase bacterial growth in the rhizosphere (Clarholm, 1985).

The presence of protozoa has been shown to increase plant growth significantly (Zwart et al., 1994). Using [^{15}N]-labelled bacteria, Kuikman et al. (1989) confirmed that grazing by protozoa increases the availability of nitrogen to plants originating from bacterial cells but also from soil organic matter. Clarholm (1985) argued that when no nitrogen is lost by exudation, the readily available carbon compounds in root exudates stimulate growth of dormant microbial populations along the root tip, resulting in mineralization of nitrogen from the soil organic matter and subsequent incorporation of nitrogen into microbial biomass. She assumed that the uptake of nitrogen from soil organic matter by rhizosphere bacteria and the subsequent release of these nutrients from bacteria due to grazing by protozoa and nematodes are of significant importance for plant growth. However, there is evidence that the microbial loop is insufficient for explaining the protozoan-mediated stimulation of plant growth. By including the amount of nitrogen lost through exudation and modeling N transformations in the rhizosphere, Robinson et al. (1989) and Griffiths and Robinson (1992) concluded that rhizosphere bacteria do not use plant-derived carbon to mineralize soil organic nitrogen to any great extent.

Microfaunal grazing may also affect plant growth by grazing-mediated changes in the composition of the microbial community (Burr and Caesar, 1984; Griffiths et al., 1999). The stimulation of plant growth in the presence of protozoa may be due to indirect effects since protozoan grazing on the rhizosphere microflora results in significant changes in the composition of the microbial community. This was confirmed in a soil system (Griffiths et al., 1999) by studying the effects of bacterial feeding nematodes, protozoa or a combination of nematodes and protozoa on the structure of the rhizosphere microbial community. The extraction of bacterial DNA revealed that protozoa changed the composition of the bacterial population in the rhizosphere, and phospholipid fatty acids showed that Gram-positive bacteria were increased while Gram-negative bacteria were decreased in treatments with protozoa. These changes were linked with distinct shifts in the metabolic diversity of the microbes assessed by Biolog®. Thus, there is increasing evidence that microbial grazers, particularly protozoa, are able to alter the composition and the functioning of the rhizosphere microbial community. Hence, grazing by protozoa may alter the composition of the rhizosphere microflora from unfavorable microorganisms towards a plant growth-promoting rhizosphere community (Burr and Caesar, 1984).

Additionally, indirect effects of fauna on plant growth are in dispersal of rhizosphere bacteria, which is supported by Knox et al. (2003), who suggest that, for example, nematodes are important vectors for colonization of rhizosphere bacteria. They used a sand-based microcosm system to investigate the ability of three species of nematodes (*Caenorhabditis elegans*, *Acrobeloides thornei* and *Cruznema* sp.) to promote rhizosphere colonization by four strains of beneficial rhizobacteria. In nearly all cases, rhizosphere colonization was substantially increased by the presence of nematodes, irrespective of bacterial or nematode species.

2.4. Microbial–Microbial Indirect Interactions

Soil microbial populations are multispecific communities, which fluctuate according to the environmental changes. In nature, every microbial cell is likely to interact with other cells, and the dynamics of every microbial population is partly a result of its interactions with other populations. Microbial life and microbial interactions are therefore inseparable, and the latter are as ubiquitous as the former. The interactions can be divided into *direct interaction* (the most important type of direct microbe-microbe interactions in the rhizosphere is antagonism, although this may simply reflect the fact that in general such activity is easier to detect and investigate than the other major types of interaction, such as mutualism and competition) (Curl and Truelove, 1986), and/or *interactions mediated by a resource* (e.g. competition for a nutrient), and/or *indirect interactions* (e.g. a strong competitor negatively affecting the predator of a weaker competitor by decreasing the food base of this predator) (Epstein, 2003).

Microorganisms are essential food chain links between primary production and soil animals. They are consumed by microscopic animals that again serve as food for larger animals. A large part of the primary production (in many ecological systems it is by far the largest fraction) is not consumed directly by soil animals. Rather, the organic material is degraded by microorganisms (Schlegel, 1993; Pace, 1997). There are several reasons for this: animals are in general incapable of hydrolyzing most polymeric plant compounds and most plant tissue is very poor in essential nutrients, e.g. nitrogen and phosphorus. Finally, animals are not capable of utilizing dilute solutions of organics, nor the (energy-containing) end products of anaerobic degradation such as methane or hydrogen sulfide. Conversely, microorganisms are capable of degrading all naturally occurring polymers and they can assimilate dissolved inorganic nutrients from the environment (Pace, 1997). They are extremely efficient in assimilating dissolved organic material, converting it into particulate food that is available to phagotrophs (Pace, 1999). Also, specialized chemolithotrophic microorganisms can harvest the energy of, for example, dissolved sulfide to produce their cells (Schlegel, 1993). Microorganisms thus play an essential role as food chain links between primary production and animals.

3. Indirect Interactions in The Rhizosphere of Different Types of Grasslands

Grasslands can be classified into tropical, flooded, temprate, tundra-steppe and desert that are vary in nutrients content, pH, organic matter, amounts of light, water and heat they obtained leading to formation of specific soil organisms communities adapted to the particular environment. These differences are not always obvious to distinguish and in terms of biotic environment temprate grassland can gradualy change, for example, into flooded or desert grasslands depending on amounts of heat and moisture available.

Humidity excess in the flooded grasslands favours the maintenance of the species adapted for humidity excess as rushes and sedges. Legumes in flooded grasslands are missing in general or contribute to the vegetation cover with 0.2-0.8% (Moisuc et al., 2010). High moister conditions are likely defines not only which plant species will inhabit such environment (Beumer et al., 2008) but other soil organisms too, affecting an outcome of biotic interactions. Flooded grasslands might consist of anaerobic zones with soil, which favors anaerobically growing bacteria such as methanogens and methanotrophs that still remain viable in soil under unfavourable (aerobic) conditions. Even though methane transfer from the soil to the atmosphere occurs mostly through the aerenchyma of plants, but ca 60-90% of methane produced in the anaerobic zones of grasslands is reoxidised in their aerobic zones – rhizosphere (Le Mer and Roger, 2001). Additionally, the rhizosphere supports the activity of microbial communities that decompose organic matter and this microbial decomposition is greatly influenced by the activity of soil macrofauna, especially earthworms (Brown and Doube, 2004). Due to high moister and/or anaerobic environment the increase in diversity and abundance of earthworms is restricted. Nevertheless, the activity of microbial communities is higher in soil of flooded grasslands due to high organic matter content in comparison with other types of grasslands (Ivask et al., 2007). This can be explained by the presence of semi aquatic earthworm species that are adapted to high moisture and organic matter content (Ivask et al., 2007). In general, there are no typical "flooded grassland species" in soil macrofauna, but only tolerant to high moisture content species (Plum, 2005). Regarding microfauna there is a scarce amount of information that does not allow to evaluate the role of microfauna in biotic interactions of the flooded grasslands. Neverethelss, it was shown that the microfauna is presented by protozoa and nematodes that inhabit aerobic microsites of the rhizosphere (Kimura et al., 1989) and which presumably can affect biomass and activity of microorganisms in soil.

Steppe plants are greatly depended on and helped by mycorrhizae to grow, increasing dry matter and phosphorus content of roots and phosphorus absorption rate of the plants, thus, indirectly, improving water stability of soil aggregates and microbial flora in the rhizosphere (Cai et al., 2007). Steppe microbial flora is commonly represented by *Pseudomonas* and actynomycetes that are also typical in the rhizosphere of temprate grasslands (Norovsuren et al., 2007). Soil nematode communities were investigated in semi-natural steppe grasslands of Austria and altogether 58 nematode genera were detected. Nematode diversity indices were characteristic for temperate grassland soils, but did not separate communities well (Zolda, 2008). This indicates that steppe grasslands are very similar to temprate grasslands and consequently biotic interactions will be similar too, including indirect effects between soil organisms and plants.

In spite of drought and harsh conditions, desert grasslands are not different from other grassland types regarding the presence of mycorrhizae. In desert area mycorrhizae still helps plants to resist drought and other unfavorable conditions coming along with high desert temperature (Al-Whaibi, 2009). Besides direct help in nutrients acquisition, presence of mycorrhizae indirectly enhances sand fixation ability of the host plants (Shan et al., 2009). It is also shown that fertility of desert soils is supported by carbon and nitrogen exchanges through a symbiotic fungal network dominated by dark-septate fungi *Pleosporales* (Green et al., 2008). Additionally, other rhizosphere microorganisms are also likely to positively affect the growth of desert plants. It was shown that rhizospheric microbial inoculum alone and together with fertilizers clearly enhanced growth and survival of the native desert plants *Nitraria retusa* (Baroon et al., 2009). Such enhacing effects on plants can be due to microbial aquisition of nutrients: assimilation of atmospheric nitrogen (Chowdhury et al., 2009), phosphorus solubilization by pseudomonads (Gulati et al., 2008), bacterial autotrophic nitrification and fungal nitrifier denitrification that dominate nitrogen transformation in desert soil (Crenshaw et al., 2008).

In turn different desert hylophyte plants affect rhizospheric microbial communities through changes in soil nutrient status and soil enzyme activity (Yi et al., 2009). These changes are most likely due to different plant ecophysiological characteristics that play an important role in the temporal and spatial distribution of abiotic conditions and microbial community levels. For example, two desert plants, *Atriplex halimus* that creates 'islands of salinity' in its rhizosphere system and the relatively high percentage of organic matter beneath its canopy, and the *Hammada scoparia* that has a leafless stem with a relatively lower amount of organic matter accumulation beneath its canopy, differently affected microbial community levels of bacteria and fungi in desert grassland (Barness et al., 2009). Apparently, various plant ecophysiological characteristics, including the allelopathic ability to dominate in relationships with other plants or typical dry desert salt-resistant plants, will differently affect not only microbial community but soil animals communities too (Pen-Mouratov et al., 2008).

4. Methods Used For the Study of the Rhizosphere Interactions

A broad range of approaches has been used to investigate soil food webs. Recent developments in different methods are shedding some light on rhizosphere interactions. The increasing application of molecular techniques to study rhizosphere interactions, provide the basis for sensitive broad-scale (community level) to fine-scale (species level) investigations. Such studies, when coupled with use of well characterized experimental systems, should provide important complementary information to studies using more traditional techniques.

4.1. Community Level Analysis

A logical extension to traditional isolation techniques is to investigate rhizosphere community carbon utilization profiles through application of the Biolog® system (Garland and Mills, 1991). Previously Biolog® has been used to characterize differences in microbial communities between contrasting habitat types, soil types (Zak et al., 1994). The Biolog® assay uses microtiter plates consisting of 96 wells containing separate sole carbon sources and a redox indicator dye, producing patterns of potential carbon source utilization for microbial communities. However, the main problem associated with the use of Biolog® is that, in common with all methods dependent on growth of microbial populations in artificial media, albeit with a broad range of potential substrates, it is selective (for culturable organisms) and biased toward faster growing components (Paterson et al., 1997).

To overcome this problem are used the broad-scale methods, such as phospholipid fatty acid analysis (PLFA) and DNA hybridisations and re-association kinetics, which provided culture-independent analyses of the microbial diversity in the rhizosphere. PLFA, for example, provides the broad number of bacterial taxas present in the samples (Zelles, 1997). Combinations of Biolog® and PLFA techniques have shown differences in microbial community composition of bulk and rhizospheric soils. However, these methods cannot identify certain microbial species at the community level (Soderberg et al., 2004).

4.2. Individual Level Analysis

To avoid these limitations the use of rDNA gene analysis has become promising. Polymerase chain reaction (PCR) amplification of rDNA genes from environmental DNA samples, combined with fingerprinting techniques, such as denaturing gradient gel electrophoresis (DGGE), terminal restriction fragment analysis (T-RFLP), amplified rDNA restriction analysis (ARDRA), cloning and sequencing, including pyrosequencing, provide detailed information about the species composition of whole communities (Torsvik and Ovreas, 2002; Nicol et al., 2003; Lim et al., 2010).

Such novel method as fluorescence *in situ* hybridisation (FISH) allows a phylogenetic identification of uncultured bacteria in natural environments using fluorescent specific phylogenetic probes (targeting rRNA) and fluorescence microscopy. Combining FISH with microautoradiography, which is a powerful tool to link the uptake of radiochemicals by individual cells, it can be used to detect and quantify the active population utilizing a specific substrate (Gray et al., 2000).

4.3. Detecting Functional Activity

A major advance in detecting functional activity in a soil microbial community came with the development of stable isotope probing (SIP), which involves tracking of a stable isotope atom from a particular substrate into components of microbial cells that provide phylogenetic and functional information, such as lipids, DNA, RNA. SIP has generated a lot of interest for the study of carbon flow and plant-microbial interactions. Butler et al. (2003) reported the use of PLFA-SIP to reveal the spatial and temporal difference in microorganisms utilizing root exudates in the rhizosphere of ryegrass. The use of reporter genes, such as the *lux* genes for bioluminescence, can be employed to quantify the activity of rhizobacteria around roots (Prosser et al., 1996). Light output from *lux*-marked rhizobacteria such as *Pseudomonas fluorescens* is proportional to metabolic activity (reduction of NADH) (Meikle et al., 1992) and can therefore report directly on carbon flow through the cell. To quantify rhizosphere carbon flow, and consequently environmental impacts on carbon fluxes through soil, the stable isotopic techniques are used.

4.4. Isotope Techniques

Isotopes of carbon (^{13}C and ^{14}C) have been used as tracers of carbon flow in plant-soil systems. A variety of approaches has been used and each has particular advantages in relation to evaluating impacts of elevated CO_2 on terrestrial ecosystems (Paterson et al., 1997).

$^{14}CO_2$ continuous labelling. Continuous labelling with $^{14}CO_2$ involves exposure of plants to an atmosphere containing a constant specific activity of $^{14}CO_2$ throughout the growth of the plant. Such approach allows uniform labelling of all plant carbon pools and therefore uniform labelling of both chemically simple and complex rhizodeposits. Therefore, continuous labelling facilitates the quantification of total carbon inputs to soil throughout the growth period of the plant. This approach is appropriate to study gross changes in the rhizodeposition as a consequence of environmental parameters (Paterson et al., 1997).

Pulse-labelling approaches. Pulse-labelling has the advantage that the methodologies involved are simpler and cheaper than for continuous labelling. Pulse-labelling of plants with $^{14}CO_2$, followed by FISH-microautoradiography analysis of the rhizosphere soil of the plant, could reveal the identity of bacteria utilizing root exudates (Singh et al., 2004).

^{13}C techniques. Relatively recently, ^{13}C analysis of organic components has been used to study carbon flow in terrestrial ecosystems (Balesdent and Balabane, 1992). ^{13}C can be used as a tracer in the same way as ^{14}C however it has been more widely used in natural abundance studies. The use of the stable isotope ^{13}C, both with labelling (Ladygina et al., 2010) and natural abundance methodologies (Ladygina et al., 2008), offers a powerful alternative to more traditional ^{14}C approaches.

No single method will elucidate all interactions, but combinations can help to answer some more specific questions. For example, the presence of particular microorganisms and their roles in the rhizosphere samples can be determined by FISH-microautoradiography (Gray et al., 2000). Then extracted nucleic acids from the rhizosphere sample can be used for isotopic arrays or for separation of labelled DNA/RNA by centrifugation. Heavy DNA/RNA can be used for construction of bacterial artificial chromosome (BAC) libraries, metagenomics or in polymerase chain reaction (PCR) to identify the taxa, which have utilized, for example, root exudates (Manefield et al., 2002).

4.5. Perturbation Experiments

To study grassland soil food webs so called perturbation experiments can be applied to detect direct and indirect effects in the rhizosphere. There are pulse (Bender, 1984) and press perturbation experimental approaches (Bender, 1984; Yodzis, 1988), which have a very important difference between them two: pulse experiments gain information on direct effects, while press perturbation experiments obtained data about both direct and indirect effects in the rhizosphere (Bender, 1984). A pulse perturbation is a relatively instantaneous alteration, while a press perturbation is a sustained alteration of population densities that acts to a higher or lower level so that a persistent change will influence the resident population or community (Yodzis, 1988). These types of experiments have a great potential in identification of direction and complexity of indirect effects and were recently applied to study grassland soil food webs (Ladygina et al., 2008).

CONCLUSION

By considering indirect interactions, a more complete understanding of the community structure of the grassland rhizosphere can be gained than if we consider direct effects only. For many years ecologists have viewed soil organisms and plants as relatively independent from each other. To unravel the complex biotic interactions in the rhizosphere, further research on rhizosphere processes requires a multidisciplinary approach and an interdisciplinary exchange of knowledge between plant physiologists, soil scientists, microbiologists and zoologists leading to a deeper understanding of rhizosphere processes that regulate mineralization, nutrient cycling and plant growth. Moreover, the differences in plant growth, plant community composition can only be understood in relation to indirect microbial-faunal, plant-microbial, faunal-plant and microbial-microbial interactions in the rhizosphere.

REFERENCES

Alphei J, Bonkowski M, Scheu S. 1996. Protozoa, Nematoda and Lumbricidae in the rhizosphere of *Hordelymus europaeus* (Poaceae): Faunal interactions, response of microorganisms and effects on plant growth. *Oecologia*, 106: 111–126.

Al-Whaibi MH. 2009. Desert plants and mycorrhizae (a mini-review). *Journal of Pure and Applied Microbiology*, 3: 457–466.

Bais HP, Park SW, Weir TL, Callaway RM, Vivanco JM. 2004. How plants communicate using the underground information superhighway. *Trends in Plant Sciences*, 9: 26–32.

Balesdent J, Balabane M. 1992. Maize root-derived soil organic carbon estimated by natural ^{13}C abundance. *Soil Biology and Biochemistry*, 24: 97–101.

Bamforth SS. 1985. The role of protozoa in litters and soil. *Journal of Protozoology*, 32: 404–409.

Bardgett RD, Wardle D, Yeates GW. 1998. Linking above-ground and below-ground interactions: how plant responses to foliar herbivory influence soil organisms. *Soil Biology and Biochemistry*, 30: 1867–1878.

Barea JM, Azcon R, Azcon-Aguilar C. 2004. Mycorrhizal fungi and plant growth promoting rhizobacteria. In: Varma A, Abbott L, Werner D, Hampp R, eds. *Plant surface microbiology*. Heidelberg, Germany: Springer-Verlag, pp. 351–371.

Barness G, Zaragoza S, Shmueli I, Steinberger Y. 2009. Vertical distribution of a soil microbial community as affected by plant ecophysiological adaptation in a desert system. *Microbial Ecology*, 57: 36–49.

Baroon Z, Yateem A, Al-Surrayai T. 2009. Enhancement of *Nitraria retusa* growth by rhizospheric microbial inoculums. Deserts Plants, 25: 21–26.

Baudoin E, Benizri E, Guckert A. 2003. Impact of artificial root exudates on the bacterial community structure in bulk soil and maize rhizosphere. *Soil Biology and Biochemistry*, 35: 1183–1192.

Bender EA. 1984. Perturbation experiments in community ecology: theory and practice. *Ecology*, 65: 1–13.

Beumer V, van Wirdum G, Beltman B, Griffioen J, Grootjans AP, Verhoeven JTA. 2008. Geochemistry and flooding as determining factors of plant species composition in Dutch-winter flooded riverine grasslands. *Science of the Total Environment*, 402: 70–81.

Bever JD. 2003. Soil community feedback and the coexistence of competitors: conceptual framework and empirical tests. *New Phytologist*, 157: 465–473.

Bonkowski M, Cheng W, Griffiths BS, Alphei J, Scheu S. 2000. Microbial-faunal interactions in the rhizosphere and effects on plant growth. *European Journal of Soil Biology*, 36: 135–147.

Briones AM, Okabe S, Umemiya Y, Ramsing NB, Reichardt W, Okuyama H. 2003. Ammonia-oxidizing bacteria on root biofilms and their possible contribution to N use efficiency of different rice cultivars. *Plant and Soil*, 259: 335–348.

Brown G, Doube BM. 2004. Functional interactions between earthworms, microorganisms, organic matter, and plants, in: C.A. Edwards (Ed.), *Earthworm Ecology*, second ed., CRC Press, pp. 213–239.

Bukovinszky T, Van Veen FJF, JongemaY, Dicke M. 2008. Direct and indirect effects of resource quality on food web structure. Science, 319: 804–807.

Burr TJ, Caesar A. 1984. Beneficial plant bacteria. *Critical Reviews in Plant Sciences*, 2: 1–20.

Butler JL, Williams MA, Bottomley PJ, Myrold DD. 2003. Microbial community dynamics associated with rhizosphere carbon flow. *Applied and Environmental Microbiology*, 69: 6793–6800.

Cai XiaoBu, Feng Gu, Qian Cheng, Gai JingPing. 2007. Effects of AM fungi on steppe plants and soil environment in Tibet plateau. *Acta Pedologica Sinica*, 44: 63–72.

Campbell R, Greaves MP. 1990. Anatomy and community structure of the rhizosphere. In: J.M Lynch, Editor. *The rhizosphere* (Wiley series in ecological and applied microbiology), A Wiley-Interscience publication, pp. 100–130.

Chowdhury SP, Schmid M, Hartmann A, Tripathi AK. 2009. Diversity of 16S-rRNA and *nifH* genes derived from rhizosphere soil and roots of an endemic drought tolerant grass, *Lasiurus sindicus*. *Europian Journal of Soil Biology*, 45: 114–122.

Clarholm M. 1985. Interactions of bacteria, protozoa and plants leading to mineralization of soil nitrogen. *Soil Biology and Biochemistry*, 17: 181–187.

Clarholm M. 1994. The microbial loop in soil. In: K Ritz, J Dighton and K.E Giller, Editors, *Beyond the Biomass*, Wiley-Sayce, Chichester, pp. 221–230.

Cocking EC. 2003. Endophytic colonisation of plant roots by nitrogen-fixing bacteria. *Plant and Soil*, 252: 169–175.

Crenshaw CL, Lauber C, Sinsabaugh RL, Stavely LK. 2008. Fungal control of nitrous oxide production in semiarid grassland. *Biogeochemistry*, 87: 17–27.

Curl EA, Truelove B. 1986. Root exudation. In: *The rhizosphere*, Springer-Verlag Berlin Heidelberg, Germany, pp. 52–54.

Duineveld BM, Kowalchuk GA, Keijzer A, van Elsas JD, van Veen JA. 2001. Analysis of bacterial communities in the rhizosphere of chrysanthemum via denaturing gradient gel electroforesis of PCR-amplified 16S rRNA as well as DNA fragments coding for 16S rRNA. *Applied and Environmental Microbiology*, 67: 172–178.

Epstein SS. 2003. Microbial Interactions. *In: Nature Encyclopedia of Life Sciences*. John Wiley & Sons, Ltd: Chichester. *http://www.els. net/[doi:10.1038/npg.els.0003285].*

Garland JL, Mills AL. 1991. Classification and characterisation of heterotrophic microbial communities on the basis of patterns of community-level sole-carbon-source utilisation. *Applied and Environmental Microbiology*, 57: 2351–2359.

Gracia-Garza JA, Bailey BA, Paulitz C, Lumsden RD, Releder RD, Roberts DP. 1997. Effect of sclerotial damage of *Sclerotinia sclerotiorum* on the mycoparasitic activity of *Trichoderma hamatum*. *Biocontrol Science and Technology*, 7: 410–413.

Gramss G, Rudeschko O. 1998. Activities of oxidoreductase enzymes in tissue extracts and sterile root exudates of three crop plants, and some properties of the peroxidase component. *New Phytologist*, 138: 401–409.

Gray ND, Howarth R, Pickup RW, Jones JG, Head IM. 2000. Use of combined microautoradiography and fluorescence in situ hybridization to determine carbon metabolism in mixed natural communities of uncultured bacteria from the genus *Achromatium*. *Applied and Environmental Microbiology*, 66: 4518–4522.

Grayston SJ, Wang SQ, Campbell CD, Edwards AC. 1998. Selective influence of plant species on microbial diversity in the rhizosphere. *Soil Biology and Biochemistry*, 30: 369–378.

Grayston SJ, Griffith GS, Mawdsley JL, Campbell CD, Bardgett RD. 2001. Accounting for variability in soil microbial communities of temperate upland grassland ecosystem. *Soil Biology and Biochemistry*, 33: 533–551.

Green LE, Porras-Alfaro A, Sinsabaugh RL. 2008. Translocation of nitrogen and carbon integrates biotic crust and grass production in desert grassland. *Journal of Ecology*, 96: 1076–1085.

Griffiths B, Robinson D. 1992. Root-induced nitrogen mineralization: A nitrogen balance model. *Plant and Soil*, 139: 253–263.

Griffiths BS, Bonkowski M, Dobson G, Caul S. 1999. Changes in soil microbial community structure in the presence of microbial-feeding nematodes and protozoa. *Pedobiologia*, 43: 297–304.

Gulati A, Rahi P, Vyas P. 2008. Characterization of phosphate-solubilizing fluorescent pseudomonads from the rhizosphere of seabuckthorn growing in the cold deserts of Himalayas. *Current Microbiology*, 56: 73–79.

Hamilton WE, Sillman DY. 1989. Effects of earthworm middens on the distribution of soil microarthropods. *Biology and Fertility of Soils*, 8: 279–284.

Ingham ER, Trofymow JA, Ames RN, Hunt HW, Morley CR, Moore JC, Coleman DC. 1986. Trophic interactions and nitrogen cycling in a semi-arid grassland soil. 1. Seasonal dynamics of the natural populations, their interactions and effects on nitrogen cycling. *Journal of Applied Ecology*. 23: 597–614.

Ivask M, Truu J, Kuu A, Truu M, Leito A. 2007. Earthworm communities of flooded grasslands in Matsalu, Estonia. *European Journal of Soil Biology*, 43: 71–76.

Jaffee BA. 1992. Population biology and biological control of nematodes. *Canadian Journal of Microbiology*, 38: 359–364.

Jaffee BA, Bastow JL, Strong DR. 2007. Supression of nematodes in a coastal grassland soil. *Biology and Fertility of Soils*, 44: 19–26.

Jentschke G, Bonkowski M, Godbold DL, Scheu S. 1995. Soil protozoa and plant growth: Non-nutritional effects and interaction with mycorrhizas. *Biology and Fertility of Soils*, 20: 263–269.

Kennedy AC. 1998. The rhizosphere and spermosphere. In: Sylvia DM, Fuhrmann JJ, Hartel PG, Zuberer DA, eds. *Principles and applications of soil microbiology*. Upper Saddle River, New Jersey: Prentice Hall, pp. 389–407.

Kimura M, Murakami H, Wada H. 1989. Survival strategy of microfauna in a submerget paddy soil. *Soil Science of Plant Nutrition*, 35: 1–7.

Knox OGG, Killham K, Mullins CE, Wilson MJ. 2003. Nematode-enhanced microbial colonization of the wheat rhizosphere. *FEMS Microbiology Letters*, 225: 227–233.

Kuikman PJ, Van Vuuren MMI, Van Veen JA. 1989. Effect of soil moisture regime on predation by Protozoa of bacterial biomass and the release of bacterial nitrogen. *Agriculture, Ecosystems and Environment*, 27: 271–279.

Ladygina N, Hedlund K. 2010. Plant species influence on microbial diversity and carbon allocation in the rhizosphere. *Soil Biology and Biochemistry*, 42: 162–168.

Ladygina N, Caruso T, Hedlund K. 2008. Dietary switching of collembola in grassland soil food webs. *Soil Biology and Biochemistry*, 40: 2898–2903.

Le Mer J, Roger P. 2001. Production, oxidation, emission and consumption of methane by soils: A review. *Europian Journal of Soil Biology*, 37: 25–50.

Lim YW, Kim BK, Kim C, Jung HS, Kim BS, Lee JH, Chun J. 2010. Assessment of soil fungal communities using pyrosequencing. *Journal of Microbiology*, 48: 284–289.

Loneragan JF. 1997. Plant nutrition in the 20th and perspectives for the 21st century. *Plant and Soil*, 196: 163–174.

Lynch JM. 1983. Plant growth regulators and phytotoxins from micro-organisms. In: JM Lynch, Editor, *Soil Biotechnology*, Blackwell Scientific Publications, London, pp. 107–120.

Lynch JM. 1990. Microbial metabolites. In: JM Lynch, Editor. *The rhizosphere* (Wiley series in ecological and applied microbiology), A Wiley-Interscience publication pp. 177–207.

Lynch JM, Whipps JM. 1990. Substrate flow in the rhizosphere. *Plant and Soil*, 129: 1–10.

Maden E, Alexander M. 1982. Transport of *Rhizobium* and *Pseudomonas* through soil. *Journal of Soil Science Society of America*, 42: 557–560.

Manefield M, Whiteley AS, Griffiths RI, Bailey MJ. 2002. RNA stable isotope probing, a novel means of linking microbial community function to phylogeny. *Applied and Environmental Microbiology*, 68: 5367–5373.

Maraun M, Migge S, Schaefer M, Scheu S. 1998. Selection of microfungal food by six oribatid mite species (Oribatida, Acari) from two different beech forests. *Pedobiologia*, 42: 232–240.

Meikle A, Killham K, Prosser JI, Glover LA. 1992. Luminometric measurement of population activity of genetically modified *Pseudomonas fluorescens* in the soil. *FEMS Microbiology Letters*, 99: 217–220.

Menge BA. 1995. Indirect effects in marine rocky intertidal interaction webs: patterns and importance. *Ecological Monographs*, 65: 21–74.

Mitra S, Aulakh MS, Wassmann R, Olk DC. 2005. Triggering of methane production in rice soils by root exudates: Effects of soil properties and crop management. *Journal of Soil Science Society of America*, 69: 563–570.

Moisuc A, Sarateanu V, Durau C, Pavel V. 2008. Approach on the evolution of grasslands floristic composition after flooding. *Research Journal of Agricultural Science,* 42: 508–513.

Nagahashi G, Douds DD. 2004. Isolated root caps, border cells, and mucilage from host roots stimulate hyphal branching of the arbuscular mycorrhizal fungus, *Gigaspora gigantea*. *Mycological Research*, 108: 1079–1088.

Nehl DB, Allen SJ, Brown LF. 1996. Deleterious rhizosphere bacteria: an integrating perspective. *Applied Soil Ecology*, 5: 1–20.

Newell K. 1984. Interactions between two decomposer Basidiomycetes and a collembolan under Sitka spruce: Distribution, abundance and selective grazing. *Soil Biology and Biochemistry*, 16: 227–233.

Nicol GW, Glover LA, Prosser JI. 2003. Spatial analysis of archaeal community structure in grassland soil. *Applied and Environmental Microbiology*, 69: 7420–7429.

Nikolyuk VF. 1969. Some aspects of the study of soil protozoa. *Acta Protozoologica*, 7: 99–108.

Norovsuren Zh, Zenova GM, Mosina LV. 2007. Actinomycetes in the rhizosphere of semidesert soils of Mongolia. *Pochvovedenie*, 4: 457–460 (in Russian).

Odintsova TI, Rogozhin EA, Baranov Y, Musolyamov AKh, Yalpani N, Egorov TA, Grishin EV. 2008. Seed defensins of barnyard grass *Echinochloa crusgalli* (L.) Beauv. *Biochimie*, 90: 1667–1673.

Ohgushi T. 2008. Herbivore-induced indirect interaction webs on terrestrial plants: the importance of non-trophic, indirect, and facilitative interactions. *Entomologia Experimentalis et Applicata*, 128: 217–229.

Pace NR. 1997. A molecular view of microbial diversity and the biosphere. *Science*, 276: 734–740.

Pace NR. 1999. Microbial ecology and diversity. *ASM News*, 65: 328–333.

Paterson E, Hall JM, Rattray EAS, Griffiths BS, Ritz K, Killham K. 1997. Effect of elevated CO_2 on rhizosphere carbon flow and soil microbial processes. *Global Change Biology*, 3: 363–377.

Pen-Mouratov S, Barness G, Steinberger Y. 2008. Effect of desert plant ecophysiological adaptation on soil nematode communities. *Europian Journal of Soil Biology*, 44: 298–308.

Plum N. 2005. Terrestrial invertebrates in flooded grassland: a literature review. *Wetlands*, 25: 721–737.

Prosser JI, Killham K, Glover LA, Rattray EAS. 1996. Luminescence-based systems for detection of bacteria in the environment. *Critical Reviews in Biothechnology*, 16: 157–183.

Robinson D, Griffiths BS, Ritz K, Wheatley R. 1989. Root induced nitrogen mineralization: A theoretical analysis. *Plant and Soil*, 117: 185–193.

Schlegel HG. 1993. *General Microbiology*. 7th edn. Cambridge, UK: Cambridge University Press.

Schmidt O, Doube BM, Ryder MH, Killham K. 1997. Population dynamics of *Pseudomonas corrugata* 2140R Lux 8 in earthworm food and in earthworm casts. *Soil Biology and Biochemistry*, 29: 523–528.

Shan BQ, He XL, Duan XY. 2009. Diversity and spatial distribution of arbuscular mycorrhiza fungi of phalanx clonal plant in the Mu Us desert. *Acta Prataculturae Sinica*, 18: 146–154.

Singh BK, Millard P, Whiteley AS, Murrell JC. 2004. Unraveling rhizosphere-microbial interactions: opportunities and limitations. *Trends in Microbiology*, 12: 386–393.

Smalla K, Wieland G, Buchner A, Zock A, Parzy J, Kaiser S, Roskot N, Heuer H, Berg G. 2001. Bulk and rhizosphere soil bacterial community studied by denaturing gradient gel electrophoresis: plant dependent enrichment and seasonal shift revealed. *Applied and Environmental Microbiology*, 67: 4742–4751.

Soderberg KH, Probanza A, Jumpponen A, Baath E. 2004. The microbial community in the rhizosphere determined by community level-physiological profile (CLPP) and direct soil- and cfu-PLFA techniques. *Applied Soil Ecology*, 25: 135–145.

Somers E, Vanderleyden J, Srinivasan M. 2004. Rhizosphere bacterial signalling: a love parade beneath our feet. *Critical Reviews in Microbiology*, 30: 205–240.

Strauss SY. 1991. Indirect effects in community ecology: their definition, study and importance. *Trends in Ecology and Evolution*, 6: 206–210.

Thorpe I, Killham K, Prosser JI, Glover LA. 1993. Novel method for the study of the population dynamics of a genetically modified micro-organism in the gut of the earthworm *Lumbricus terrestris*. *Biology and Fertility of Soils*, 15: 55–59.

Toal ME, Yeomans C, Killham K, Meharg AA. 2000. A review of rhizosphere carbon flow modelling. *Plant and Soil*, 222: 263–281.

Torsvik V, Ovreas L. 2002. Microbial diversity and function in soil: from genes to ecosystems. *Current Opinion in Microbiology*, 5: 240–245.

Trolove SN, Hedley MJ, Kirk GJD, Bolan NS, Loganathan P. 2003. Progress in selected areas of rhizosphere on P acquisition. *Australian Journal of Soil Research*, 41: 471–499.

Tylka GL, Hussey RS, Roncadori RW. 1991. Interactions of vesicular-arbuscular mycorrhizal fungi, phosphorus, and *Heterodera glycines* on soybean. *Journal of Nematology*, 23: 122–133.

Van der Weerden NL, Lay FT, Anderson MA. 2008. The plant defensin, NaD1, enters the cytoplasm of Fusarium oxysporum hyphae. *Journal of Biological Chemistry*, 283: 14445–14452.

Wardle DA, Yeates GW. 1993. The dual importance of competition and predation as regulatory forces in terrestrial ecosystems: evidence from decomposer food-webs. *Oecologia*, 93: 393–306.

Wardle DA, Bardgett RD, Kliromonas JN, Setala H, Wim H. van der Putten, Wall DH. 2004. Ecological limkages between aboveground and belowground biota. *Science*, 304: 1629–1633.

Werner D. 1998. Organic signals between plants and microorganisms. In: Pinton R, Varanini Z, Nannipieri P, eds. *The rhizosphere: biochemistry and organic substances at the soil-plant interfaces*. New York: Marcel Dekker Inc., pp. 197–222.

Whipps JM, Budge SP. 1993. Transmission of the mycoparasite *Coniothyrium minitans* by collembolan *Folsomia candida* (Collembola: Entomobryidae) and glasshouse sciarid *Bradysia* sp. (Diptera: Sciaridae). *Annals of Applied Biology*, 123: 165–171.

White C, Tardif JC, Adkins A, Staniforth R. 2005. Functional diversity of microbial communities in the mixed boreal plain forest of central Canada. *Soil Biology and Biochemistry*, 37: 1359–1372.

Williams RH, Whipps JM, Cooke RC. 1998a. Role of soil mesofauna in dispersal of *Coniothyrium minitans*: transmission to sclerotia of *Sclerotinia scleotiorum*. *Soil Biology and Biochemistry*, 30: 1929–1935.

Williams RH, Whipps JM, Cooke RC. 1998b. Role of soil mesofauna in dispersal of *Coniothyrium minitans*: mechanisms of transmission. *Soil Biology and Biochemistry*, 30: 1937–1945.

Wipat A, Harwood CR. 1999. The *Bacillus subtilis* genome sequence: the molecular blueprint of a soil bacterium. *FEMS Microbiology Ecology*, 28: 1–9.

Yang C-H, Crowley DE. 2000. Rhizosphere microbial community structure in relation to root location and plant iron nutritional status. *Applied and Environmental Microbiology*, 66: 345–351.

Yi LP, Ma J, Li Y. 2009. Dynamics of soil enzyme activity in the rhizosphere of desert hylophyte. *Chinese Journal of Eco-Agriculture*, 17: 500–505.

Yodzis P. 1988. The indeterminacy of ecological interactions as perceived through perturbation experiments. *Ecology*, 69: 508–515.

Zahir ZA, Arshad M, Frankenberger WT. 2004. Plant growth promoting rhizobacteria: applications and perspectives in agriculture. *Advances in Agronomy*, 81: 97–168.

Zak JC, Willig MR, Moorehead DL, Wildman HG. 1994. Functional diversity of microbial communities: a quantitative approach. *Soil Biology and Biochemistry*, 26: 1101–1108.

Zelles L. 1997. Phospholipid fatty acid profiles in selected members of soil microbial communities. Chemosphere, 35: 275–294.

Zolda P. 2008. *Ecological analysis of free living soil nematode communities (roundworms) in the National Park Neusiedl* - Seewinkel (Austria). Abhandlungen der Zoologisch-Botanischen Gesellschaft in Oesterreich, 37: 85–92 (in German).

Zwart KB, Kuikman PJ, Van Veen JA. 1994. Rhizosphere protozoa: their significance in nutrient dynamics. In: J.F Darbyshire, Editor, *Soil Protozoa*, CAB International, Wallingford, pp. 93–122.

In: Grasslands: Types, Biodiversity and Impacts
Editor: Wen-Jun Zhang
ISBN: 978-1-61470-555-0

Chapter 3

EVALUATION OF SOIL QUALITY ON TEMPERATE GRASSLAND SOILS USING BIOLOGICAL AND BIOCHEMICAL PROPERTIES

J. Paz-Ferreiro*
Departamento de Edafología. Escuela Técnica Superior de Ingenieros Agrónomos.
Universidad Politécnica de Madrid, Madrid, Spain

ABSTRACT

Approximately one quarter of the agricultural surface is covered by grasslands. In the last decades there has been a dramatic increase in the productivity of many grasslands. Nowadays, intensive and traditionally managed grasslands coexist in many regions. Management changes have caused alterations, mainly in the nitrogen and phosphorus cycles and in the composition of the botanic and microbial communities of grassland soils. However, the impact of management changes on soil quality has seldom been evaluated. On the other hand, soil biochemical and biological properties are considered as suitable and quickly responsive parameters to estimate soil quality. My work wants to summarize the state of art of the study of soil quality in grassland soils, comparing different managed grassland and utilizing soil biochemical and biological properties as a tool to evaluate soil quality. This review will put a strong emphasis in the temperate grasslands located in NW Spain.

GRASSLAND SOILS

Grasslands occupy approximately one quarter of Earth´s land surface. They may occur naturally or they can be the result of anthropogenic activity. Native grassland generally results in areas with a hydric deficit. As an example there are the temperate grasslands which are located at mid-latitude grasslands, and including the Great Prairies of North America, the Argentinean Pampa and the Eurasian steppes. In tropical and subtropical areas flooded

* Correspondence Address: E-mail: jpaz@udc.es

grasslands (which are flooded seasonally or year-round) can occur. An example of the latter would be the Everglades in Florida the Everglades of Florida or the Pantanal in South America. Natural montane grasslands are located on high altitude areas around the globe, as could be the Páramo of the Andes Mountains or many areas in the Tibetan-Plateau. Xeric grasslands also know as desert grasslands are characterized by a strong moisture deficit and are located in areas were the precipitation is under 300 mm precipitation grasslands located in deserts and xeric shrublands ecoregions. An example would be the coastal area of Ethiopia.

On the other hand, anthropogenic grasslands can be exemplified by the tallgrass prairies in the American Midwest that have been extended eastward as a consequence of human activity.

Also anthropogenic influence of grassland can be found in intensive management systems. These have been developed in order to maximize food and fibres production in areas where extensive cattle management was common or clear-cutting areas that have been previously occupied by woodland. Intensive grassland management has been associated with water and air contamination. However, studies that assess soil quality are not so abundant.

Soil Quality

In spite of the obvious interest that soil quality represented for humankind since the beginning of agriculture, its study has been quite recent and just undertaken by soil scientist in the last 20 years. However, sustainable land use requires the maintenance or the improvement of soil quality. Possibly, the reason for this neglection was the fact that measuring soil quality is difficult compared with quantifying air or water quality. This is due to the fact that water and air quality involves measuring the presence of some contaminants above certain thresholds, while determination of soil quality requires the measurement of several physical, chemical and biological properties. Nowadays, the most commonly accepted definition of soil quality is that of Doran and Parkin (1994) which defined it as the capacity of soil to sustain biological capacity and diversity, maintain environmental quality and promote plant and animal health.

There are several constraints when studying soil quality. The first one is to find suitable properties to be used as soil quality indicators. Biological properties have become increasingly popular to measure soil quality as they respond quickly to changes in management, especially when compared to chemical or physical properties (Karlen et al., 1996). On the other hand their response can be confused by abiotic factors (Paz-Ferreiro et al., 2010) and thus show a high temporal variability which can make some properties to double or half the value in a period of just one month (Paz-Ferreiro et al., 2011). When I refer to biological properties I am meaning several microbiological and biochemical parameters including microbial biomass carbon (C_{mic}), microbial biomass nitrogen, and microbial biomass; the microbial quotient (C_{mic}/organic C ratio); basal respiration, the microbial metabolic quotient (q_{CO2}=respiration/C_{mic}) and enzyme activity, including activities linked to the size and activity of the microbial community (catalase, dehydrogenase) and hydrolitic activities involved in C, N, P or S cycling. Also fauna has been used more as bioindicators of soil quality, being nematodes, arthropods and the number and activity of earthworms the ones more commonly used. Recently, microbial diversity parameters such as community level physiological profile, phospholipid fatty acids, the ratio of Gram-negative/Gram-positive bacteria, the ratio of fungal/bacterial microorganisms, and free-living diazotrophic bacteria have been identified as important biological indicators of soil quality.

Another problem to measure soil quality is the election of a suitable reference soil. Although finding a suitable baseline is difficult and some authors even hypothesize that is not possible as natural landscapes have been too altered by humankind (Silvertown et al., 2010), several authors have used climax soils as reference for forest soils (Trasar-Cepeda et al., 1998; Armas et al., 2007; Zornoza et al., 2007; Chaer et al., 2009).

Currently, there is a great interest to assess soil quality in grassland soil due to the huge amount of land that they occupy and the ecosystem services that they provide (van Eekeren et al., 2010). However, not so many studies on the topic have been carried out to date and when studies are available, and the studies reported so far do not usually deal with the quantification of soil quality (see, for example, An et al., 2010). However, over the last 10 years many expressions been proposed to evaluate soil quality, and some of them are aimed towards grassland soils.

My review focuses on the quantification of soil quality in grassland soils, particular attention would be devoted to studies using soil biological properties as indicators of soil quality, as they are deemed to be more sensitive to changes in soil quality and to studies carried out in temperate grasslands, as they are far more abundant than those carried out in other areas.

LAND USE CONVERSION EFFECT ON SOIL QUALITY

Changing land use from one type to another type generally affects the soil ecosystems. Natural systems and distributed systems are different from their biological activities. β-glucosidase and acid phosphatase, and arylsulfatase (except α-glucosidase and β-glucosaminidase) based on land use showed the following orders, respectively: grassland > forest > agriculture (Acosta-Martinez et al. 2007). Similarly, Saviozzi et al. (2001) compared soil quality in a native grassland, a poplar forest and a agricultural soil cropped with corn. These authors measured a large amount of soil biological properties (humified organic C, total N, light fraction content and composition, water-soluble organic C, water-soluble carbohydrates, phenolic substances, microbial biomass C, soil respiration) and several enzyme activities such as alkaline phosphatase, protease, β-glucosidase, urease, catalase and dehydrogenase) and showed that native grassland exhibit a good quality level, even higher than native forest systems. Microbial activity was much higher in the forest soil and in the native grassland soil than in the agricultural field. Similar results have been found later on larger scale soil quality monitoring programs (Tischer, 2005).

In a later study comparing forest climax soils developed under oaktree and native grasslands, Paz-Ferreiro et al. (2007) showed that native grassland exhibited a faster rate of nutrient turnover than forest climax soils, especially in the nitrogen cycle and attributed these differences to the deposition of dung by grazing cattle, and to the greater rhizospheric activity of the grassland vegetation. Paz-Ferreiro et al. (2007) also found that the association among variables in native grasslands and forest climax soils is similar and probably determined by macroclimatic factors.

Sicardi et al. (2004) found lower values of several soil biological properties (soil respiration, dehydrogenase, fluorescein diacetate hydrolysis and acid and alkaline phosphatase activities) after conversion of grassland to Eucalyptus forest. On the contrary, the implementation of pasture ecosystems in forest land has been shown to decrease soil microbial biomass carbon (Lopes et al., 2010).

Grassland soils have been used as a standard of high quality soils in studies conducted in the region of the Great Plains (Carpenter-Boggs et al., 2003), where ample areas of grassland have been used as croplands. These authors found that grasslands contained approximately 160% greater labile carbon and 50% greater microbial biomass, and had approximately 150% greater dehydrogenase activity than their cropped counterparts, either under conventional grass or no-tilled management.

Recently, the BSQ index (Biological Soil Quality) has been proposed to measure soil quality (Gardi et al., 2002) as an integrated approach. The BSQ index is based on the idea that soil with higher quality will have a higher diversity of microarthropods group and that it will be specialized (thus, well adapted to particular soil habitats). The application of biological indicators is often limited by the difficulties in classifying microarthropods. However, and advantage of the BSQ index is that taxonomic analysis is not required for its calculation. The BSQ index is based in the fact that microarthropods show morphological characters that reveal adaptation to soil environments. Thus, within each higher taxon, the BSQ method requires searching for the biological form (morpho-type) that is most adapted to soil. This type will receive an eco-morphological score (EMI), proportionate to its adaptation level (Parisi et al., 2005). Eco-morphological scores range from 1 to 20. After scores are assigned to each taxon, scores for all taxons are summed up.

The index was used for the first time to compare soil quality in native grasslands in Italy with adjacent degraded agricultural land (Parisi et al., 2005). The authors found that grassland soils were hotspots of biodiversity and that the results obtained using the BSQ index as indicator of soil quality were in accordance with the results obtained using more traditional soil quality parameters such as organic matter content and aggregate stability.

Results obtained in similar experiments using earthworms as indicators of soil quality (Diddem, 2001) have found that the abundance, biomass and species diversity is significantly higher in grassland soils than in horticultural soils.

Additionally, ecological indices based on nematode ability to colonise newly created habitats have been developed and there is a growing interest on their use. The maturity index (MI; Bongers, 1990) is based on the allocation of nematode taxa (excluding plant-parasitic genera) into five groups representing different life strategies and ecological requirements. The maturity index is the median of the frequency distribution of these c-p groups in a given sample. Urzrelai et al. (2000) found similar maturity index in grassland when compared to other land uses, but they showed that the diversity of nematodes is higher is grassland compared with agricultural fields and waste dumping sites.

Grassland Soil Quality

Responses of carbon and nitrogen cycles (see, for example the work by Manley et al., 1995) to management and soil physical properties and vegetation changes due to management have been measured for quite a long time (Crawford and Middle, 1977). However, the study of grassland soil quality is more recent. Thus, Schacht et al. (1996) studied the effect of two management practices that have commonly been used in grassland soils, namely burning and mowing on two tallgrass sites located in Nebraska. Their study, as most studies afterwards was based on soil chemical and physical properties and they showed that repeated burning or mowing treatments could decrease infiltration rates.

Most of the studies on soil quality have been carried out on forest climax ecosystems (Chaer et al., 2009), polluted soils or croplands (Bastida et al., 2008) and studies are not so abundant in grassland soils and when available they have been carried out in temperate grasslands and not in other biomes. Our knowledge of the topic is really limited in tropical grasslands or montane grasslands, with the exceptions in the case of tropical grasslands of some few works that however do not consider biological properties (see, for example Wang and Gong, 1998) and of a work dealing with soil quality in montane grasslands which will be discussed in the following paragraphs (Mijangos et al., 2010). In tropical areas biotic and abiotic factors differ from temperate areas. In the case of abiotic factors, temperate grasslands receive an average of 305 to 1016 mm of rain per year, while tropical grasslands receive 635 to 1524 mm. Moreover, rainfall is not distributed evenly in both types of ecosystems through the year. Also, climate determines the growing season for grassland plants and influences nutrient cycling. There are few studies that assess how soil chemical properties vary in grassland climosequences (but see Brye et al., 2004) and almost no information exists in the case of the variation of biological properties in a climosequence of grasslands. In temperate regions plants are able to grow when the temperature is above freezing, while in the tropics, where the temperature is warm enough for plant growth throughout the year, rainfall is the crucial factor. This is of capital importance as small changes in abiotic factors can greatly affect soil biological properties (Paz-Ferreiro et al., 2010, 2011).

Bastida et al. (2006) developed the MDI with the aim of investigating soil quality in arid degraded soils. The MDI is as follows:

$$MDI = [0.89(1/(1+(a/4.87)^{-2.5})] + [0.86(1/(1+(b/11.09)^{-2.5})] + [0.84(1/(1+(c/1.79)^{-2.5})] + [0.75(1/(1+(d/95.03)^{-2.5})] + [0.72\ (1/(1+(e/18.01)^{-2.5})]$$

where a is the value of dehydrogenase activity in mg INTF g^{-1} h^{-1}; b is the value of WSCh in mg kg^{-1}; c is the value of urease activity in mmol NH_3 g^{-1} h^{-1}; d is the value of WSC in mg kg^{-1}; e is the value of respiration in mg CO_2-C kg^{-1} soil.

The authors analyzed soil grassland soils in different stages of degradation and compared them with forest soil In light of the results obtained in the MDI for the analysed soils, percentiles of 0.25, 0.50 and 0.75 were established for the different levels of degradation. In this way MDI values of 0.65, or between 0.66 and 1.54 indicated a very high or high level of degradation, respectively, while values of between 1.55 and 3.08 suggested low levels of degradation.

In climax forest soils there is an equilibrium between soil organic matter content and some of the biochemical activities (Trasar-Cepeda et al., 1998). Based on the premises that native grassland soils are also high quality soils, comparable to climax forest soils, Paz-Ferreiro et al. (2007) presented an equation relating biochemical properties and organic matter content using a stepwise procedure in native grasslands soil from Galicia, according to the following expression:

Total C = 0.764 + $2.304 \cdot 10^{-3}$·microbial biomass C + 0.936·catalase + 0.017·urease + 0.206·phosphomonoesterase

where the properties are expressed as mg kg^{-1} for microbial biomass C, mmol H_2O_2 consumed g^{-1} h^{-1} for catalase activity, μmol NH_3 g^{-1} h^{-1} for urease activity and μmol *p*-nitrophenol g^{-1} h^{-1} for phosphomonoesterase activity.

This equation was obtained from a large number of soils with an ample range of pH values and organic matter contents and thus it is believed to be highly representative. Thus, in grasslands soil quality can be estimated calculating the theoretical carbon content from Paz-Ferreiro´s et al. (2007) equation (Ct) and relating it to the actual carbon content of the soil (Cr). Native grassland soils would have theoretically a 100 Ct/Cr value of 100.

On a later study, Paz-Ferreiro et al. (2009) tried to determine the threshold values of high quality soils, in this way 100 Ct/Cr values between 85-115 % would suggest that grassland soil quality has not been altered; values lower than 85% would imply biochemical degradation, while values higher than 115% would point out a transitory state of high microbiological and biochemical activity (biochemical euphoria). This reasoning was validated on a further work conducted on intensively managed grasslands (Paz-Ferreiro et al., 2009). In a latter study, Paz-Ferreiro et al. (2011) found that the equation was robust to seasonal variations, something of capital importance and that has been overlooked by most studies.

Sun ray plots can be used to represent the response of some soil variable to management (Moreno et al., 2009), usually, this is carried out by normalizing the values of the properties included in the sun-ray plot. As an example, Mijangos et al. (2010) represented several soil properties in a sun-ray plot (dehydrogenase, urease, β-glucosidade, acid phosphatase and arylsulphatase) and obtained a soil quality index. Data plotted were normalized by dividing the value obtained for each enzyme activity by the mean value obtained for that specific activity in control plots and then multiplied by 100. A treated soil quality index (T-SQI) with two terms that takes into account the magnitude of the increment of each biochemical property (by comparing the value of the biochemical property with the control soil, first summation) and the maintenance of the evenness among the soil biochemical properties considered shown by the "reference soil" (second summation):

$$\text{T-SQI} = 10^{\log m + (\Sigma(\log n - \log m) - \Sigma|\log n - \log ref|)/n}$$

where m is the reference (mean value of control soils, fixed to 100%) and n are the measured values as percentages of the reference. Mijangos et al. (2010) evaluated the T-SQI index comparing limed grassland and control grassland soils and found no variations in the T-SQI between limed and control soils.

Puglisi et al. (2006) proposed an index of soil quality, AI3.

AI3 = 7.87 β-glucosidase – 8.22 phosphatase – 0.49 urease

where the enzymatic activities were measured as μmol product g^{-1} h^{-1}.

This index was validated using six different data sets. Lower values of the index AI3 indicated a better soil quality. Originally AI3 was developed from high quality soils and soils with a lower level of soil quality (intensive exploited agricultural soils, soils polluted from tannery and landfill effluents, mining activities and eroded soils) and not exclusively for grassland soils. Thus, the index was able highly sensitive as it was able to discriminate between control soils and soils a variety of altered soils which included heavy metal polluted soils. This index was validated by Paz-Ferreiro et al. (2009) using grasslands under different management systems, these authors found that the results obtained with AI3 and with 100 Ct/Cr were similar. This is to my knowledge the first time that two soil quality equations are compared for a same dataset of soils.

Conclusions

Native grasslands are characterized by a high quality, which can be damaged as a consequence of over-exploitation. Grassland soil quality has been not explored enough, in particularly its quantification, although it seems that conversion of grassland to other land uses or intensive grassland management can result in a deterioration of soil quality. However, more studies are necessary, in particular in grassland located in the tropical areas, montane grasslands or in flooded grasslands. Intensive grassland management in temperate areas has resulted in a decrease of soil quality. Expressions have appeared in the last years with a potential to quantify soil quality, although more research is needed in this area.

References

Acosta-Martinez V, Cruz L, Sotomayer-Ramirez D, Perez-Alegria L. 2007. Enzyme activities as affected by soil properties and land use in a tropical watershed. *Applied Soil Ecology,* 35:35–45.

An SS, Huang YM, Zheng FL. 2009. Evaluation of soil microbial indices along a revegetation chronosequence in grassland soils on the Loess Plateau, Northwest China. *Applied Soil Ecology,* 41:286-29.

Armas CM, Santana B, Mora JL, Notario JS, Arbelo CD, Rodríguez-Rodríguez A. 2007. A biological quality index for volcanic Andisols and Aridisols (Canary Islands, Spain): Variations related to the ecosystem degradation. *Science of the Total Environment,* 378:238-244.

Bastida F, Moreno JL, Hernández T, García C. 2006. Microbiological degradation index of soils in a semiarid climate. *Soil Biology and Biochemistry,* 38:3463–3473.

Bastida F, Zsolnay S, Hernández T, García C. 2008. Past, present and future of soil quality indices: A biological perspective. *Geoderma,* 147:159–171.

Bongers T. 1990. The maturity index: an ecological measure of environmental disturbance based on nematode species composition. *Oecologia,* **83**:14–19

Brye KR, West CP, Gbur EE. 2004. Soil quality differences under native tallgrass prairie across a climosequence in Arkansas. *The American Midland Naturalist,* 152:214-230.

Carpenter-Boggs L, Stahl PD, Lindstrom MJ, Schumacher TE. 2003. Soil microbial properties under permanent grass, conventional tillage, and no-fill management in South Dakota. ***Soil and Tillage Research,*** 71:15-23.

Chaer GM, Myrold DD, Bottomley PJ. 2009. A soil quality index based on the equilibrium between soil organic matter and biochemical properties of undisturbed coniferous forest soils of the Pacific Northwest. *Soil Biology and Biochemistry,* 41:822-830.

Crawford AK, Middle MJ. 1977. The effect of trampling on neutral grassland. *Biological Conservation,* 12:135-142.

Didden WAM. 2001. Earthworm communities in grasslands and horticultural soils. *Biology and Fertility of Soils,* 33:111-117.

Doran JW, Parkin TB. 1996. Quantitative indicators of soil quality: A minimum data set. p. 25–37. *In* J.W. Doran et al. (ed.) Methods for assessing soil quality. SSSA, Madison, WI.

Gardi C, Tomaselli M, Parisi V, Petraglia A, Santini C. 2002. Soil quality indicators and biodiversity in northern Italian permanent grasslands. *European Journal of Soil Biology,* 38:103-110.

Karlen DL, Gardner JC, Rosek MJ. 1998. A soil quality framework for evaluating the impact of CRP. *Journal of Production Agriculture,* 11:56-60.

Lopes MM, Salviano AAC, Araujo ASF, Nunes LAPL, Oliveira ME. 2010. Changes in soil microbial biomass and activity in different Brazilian pastures. *Spanish Journal of Agricultural Research,* 8:1253-1259.

Manley JT, Schuman GE, Reeder JD, Hart RH. 1995. Rangeland soil carbon and nitrogen responses to grazing. *Journal of Soil and Water Conservation,* 50:294-298.

Mijangos I, Albizu I, Epelde L, Amezaga I, Mendarte S, Garbisu C. 2010. Effects of liming on soil properties and plant performance of temperate mountainous grasslands. *Journal of Environmental Management,* 91:2066-2074.

Moreno B, Garcia-Rodriguez S, Cañizares R, Castro J, Benítez E. 2009. Rainfed olive farming in south-eastern Spain: Long-term effect of soil management on biological indicators of soil quality. *Agriculture, Ecosystems & Environment,* 131:333-339.

Parisi V, Menta C, Gardi C, Jacomini C, Mozzanica E. 2005. Microarthropod communities as a tool to assess soil quality and biodiversity: a new approach in Italy. *Agriculture Ecosystems and Environment,* 105:323–333.

Paz-Ferreiro J, Trasar-Cepeda C, Leirós MC, Seoane S, Gil-Sotres F. 2007. Biochemical properties of acid soils under native grassland in a temperate humid zone. *New Zealand Journal of Agricultural Research,* 50:537-548.

Paz-Ferreiro J, Trasar-Cepeda C, Leirós MC, Seoane S, Gil-Sotres F. 2009. Biochemical properties in managed grassland soils in a temperate humid zone: modifications of soil quality as a consequence of intensive grassland use. *Biology and Fertility of Soils* 45:711-722.

Paz-Ferreiro J, Trasar-Cepeda C, Leirós MC, Seoane S, Gil-Sotres F. 2010. Effect of management and climate on biochemical properties of grassland soils from Galicia (NW Spain). *European Journal of Soil Biology,* 46:136-143.

Paz-Ferreiro J, Trasar-Cepeda C, Leirós MC, Seoane S, Gil-Sotres F. 2011. Intra-annual variation in biochemical properties and the biochemical equilibrium of different grassland soils under contrasting management and climate. *Biology and Fertility of Soils* (in press).

Puglisi E, Del Re AAM, Rao MA, Gianfreda L. 2006. Development and validation of numerical indexes integrating enzyme activities of soils. *Soil Biology and Biochemistry,* 38:1673-1681.

Saviozzi A, Levi-Minzi R, Cardelli R, Rifaldi R. 2001. A comparison of soil quality in adjacent cultivated, forest and native grassland soils. *Plant and Soil,* 233:251–259.

Schacht WH, Stubbendieck J, Bragg TB, Smart AJ, Doran JW. 1996. Soil quality response of reestablished grasslands to mowing and burning. *Journal of Range Management,* 49:458-463.

Sicardi M, García-Préchac M, Frioni L. 2004. Soil microbial indicators sensitive to land use conversion from pastures to commercial *Eucalyptus grandis* (Hill ex Maiden) plantations in Uruguay. *Applied Soil Ecology,* 27:125-133

Silvertown J, Tallowin J, Stevens C, Power SA, Morgan V, Emmett B, Hester A, Grime PJ, Morecroft M, Buxton R, Poulton P, Jinks R, Bardgett RD. 2010. Environmental myopia: a diagnosis and a remedy. *Trends in Ecology and Evolution,* 25:556-561.

Tischer, S. 2005. Microbial biomass and enzyme activities on soil monitoring sites in Saxony-Anhalt, Germany. *Archives of Agronomy and Soil Science,* 51:673-685.

Trasar-Cepeda C, Leirós MC, Gil-Sotres F, Seoane S. 1998. Towards a biochemical quality index for soils: an expression relating several biological and biochemical properties. *Biology and Fertility of Soils,* 26:100–106.

Urzelai A, Hernández AJ, Pastor J. 2000. Biotic indices based on soil nematode communities for assessing soil quality in terrestrial ecosystems. *The Science of the Total Environment,* 247:253-261.

van Eekeren N, de Boer H, Hanegraaf M, Bokhorst J, Nierop D, Bloem J, Schouten T, de Goede R, Brussaard L. 2010. Ecosystem services in grassland associated with biotic and abiotic soil parameters. *Soil Biology and Biochemistry,* 42:1491-1504.

Wang X, Gong Z. 1998. Assessment and analysis of soil quality changes after eleven years of reclamation in subtropical China. *Geoderma,* 81:339–355.

Zornoza R, Mataix-Solera J, Guerrero C, Arcenegui V, García-Orenes F, Mataix-Beneyto J, Morugán A. 2007. Evaluation of soil quality using multiple lineal regression based on physical, chemical and biochemical properties. *Science of the Total Environment,* 378:233–237.

In: Grasslands: Types, Biodiversity and Impacts
Editor: Wen-Jun Zhang
ISBN: 978-1-61470-555-0

Chapter 4

ORCHARDGRASS: A VALUABLE PERENNIAL PASTURE GRASS ADAPTED TO DIFFERENT ENVIRONMENTAL CONDITIONS

Pablo Luis Peri[1]

[1] Universidad Nacional de la Patagonia Austral (UNPA)-INTA-CONICET,
CC 332 (CP 9400), Río Gallegos, Santa Cruz, Argentina.

ABSTRACT

Orchardgrass is a widespread perennial grass, which is well-adapted to dry conditions and is suitable for silvopastoral systems due to its shade tolerance. The main environmental (temperature, nitrogen, water and shade) and management (regrowth duration) factors that affect morphology, physiology, dry matter (DM) production and nutritive value (crude protein, organic matter digestibility and macro-nutrient concentrations) of orchardgrass (*Dactylis glomerata* L.) in temperate climate are reviewed. Regrowth duration is a management factor that can be modified through the frequency and severity of defoliation (e.g. infrequent cutting for hay or silage, rotational or continuous grazing). The emphasis is on open pasture and silvopastoral systems conditions. This is followed by a review of how DM production could be predicted from a canopy photosynthesis model based on the photosynthetic capacity of leaves, the light intercepted by leaf surfaces (dependent upon canopy architecture and leaf area index, (LAI)). The predictive capability physiologically based pasture models makes them powerful tools for pasture management or in assisting agronomists to improve practices in pastoral or silvopastoral systems.

Keywords: dry matter production; morphology; nutritive value; Nitrogen use efficiency (NUE); radiation use efficiency (RUE); water use efficiency (WUE).

1. Introduction

Orchardgrass (*Dactylis glomerata* L.) is a widespread perennial grass, which is well-adapted to dry conditions. For example, orchardgrass is a major pasture grass used in dryland pastoral environments in New Zealand (Moloney, 1993) being recommended for use in both flat dryland and hill country systems because of its moderate fertility requirement, persistence and drought tolerance. Also, research with widely spaced radiata pine (*Pinus radiata* D. Don) has suggested that orchardgrass is a suitable grass for silvopastoral systems in temperate climates due to its shade tolerance (Devkota et al., 1997; Peri et al., 2001a, 2002b, 2007a).

All pasture species have the potential to produce a theoretical maximum yield when grown under optimum conditions (Alberda, 1962). Under non limiting conditions in Canterbury sub-humid temperate environment, a 10 year old orchardgrass dominant pasture was shown to produce a potential maximum yield of 28.6 t DM/ha/y (Peri et al., 2002c). The potential growth recorded for the Canterbury open pastures was consistent with potential yields reported in other countries. In France, Lemaire et al. (1983) reported for orchardgrass a potential growth rate of 154 kg DM/ha/d with the application of 210 kg N/ha. In Finland, Rinne (1978) reported for orchardgrass pastures fertilised with 300 kg N/ha and grazed with dairy cows a mean DM production of 139±14 kg DM/ha/d during 34 days of regrowth. However, the productivity and nutritive value of a pasture is dependent on the interaction of environmental and management factors that control growth by modifying the photosynthetic capacity and morphological aspects of the sward (Peri et al. 2002a, 2003c, 2006, 2007a).

Therefore, the aim of this Chapter was to review the main environmental (temperature, nitrogen, water and shade) and management (regrowth duration) factors that affect morphology, physiology, dry matter (DM) production and nutritive value (crude protein, organic matter digestibility and macro-nutrient concentrations) of orchardgrass (*Dactylis glomerata* L.) in temperate climate and with emphasis on open pasture and silvopastoral systems conditions. This is followed by a review of how DM production could be predicted from a canopy photosynthesis model based on the photosynthetic capacity of leaves, the light intercepted by leaf surfaces (dependent upon canopy architecture and leaf area index, (LAI)) and partitioning of photosynthates to respiration.

2. Effect of Environmental and Management Factors on Orchardgrass Growth

2.1. Solar Radiation and Shade

For all plants, the seasonal input of solar energy as photosynthetically active radiation (PAR) is the main determinant of growth when factors such as nutrients, water and temperature are non-limiting (Monteith, 1977). In such conditions the conversion of PAR to DM is conservative among C_3 species at about 1.4 g MJ^{-1} (Sinclair and Muchow, 1999). This concept has been utilised for the development of predictive models, particularly for annual crops (Sheehy and Cooper, 1973). However, trees modify both the intensity and quality of the incoming radiation. Specifically, in silvopastoral systems understorey plants experience frequent fluctuations in irradiance from full sun to shade caused by tree canopy shading (Peri et al., 2002b). The time scale of full sunlight/shade periods is dependent on the size of the tree, crown shape, tree planting density, silvicultural practices and the development of foliage area of the trees (Miah et al., 1995; Peri et al., 2007a). Thus, as expected there is a negative linear relationship between tree population and light transmission (Anderson et al., 1978, Peri et al., 2001a, 2007a). This decrease in light transmission reduces pasture yield. For example, in Canterbury plains (New Zealand) the mean total dry matter (DM) orchardgrass production was 8.2 t DM ha^{-1} yr^{-1} in open, 7.3 t DM ha^{-1} yr^{-1} in open + wooden slats (43% of the open photosynthetic photon flux density, PPFD), 6.3 t DM ha^{-1} yr^{-1} under trees shade (58% of the open PPFD) and 3.8 t DM ha^{-1} yr^{-1} in the trees+slats treatment (24% of the open PPFD) (Peri et al., 2007a). These differences in DM production were driven by DM growth rates (figure 1) being lower under trees and trees+slats compared with the full sunlight treatment in all seasons. The mean DM production rate of orchardgrass pasture for the grazing seasons (September-April) for two years was 30 kg DM ha^{-1} d^{-1} in open, 26 kg DM ha^{-1} d^{-1} in open+slats, 21 kg DM ha^{-1} d^{-1} under trees and 14 kg DM ha^{-1} d^{-1} under trees+slats. Joshi et al. (1999) reported that yield of irrigated orchardgrass pasture under 650 trees/ha (18% of the open PPFD) was reduced by 55% and by 16% under 300 trees/ha (40% of the open PPFD) compared with open pasture. Similarly, in northern Greece, Braziotis and Papanastasis (1995) reported that orchardgrass DM production during spring was reduced by 55% under a 20-year-old maritime pine (*Pinus pinaster* Aiton.) plantation thinned at 1750 trees/ha (mean light intensity of 31% of the open area) compared with pastures under 1000 trees/ha (mean light intensity of 41% of the open area).

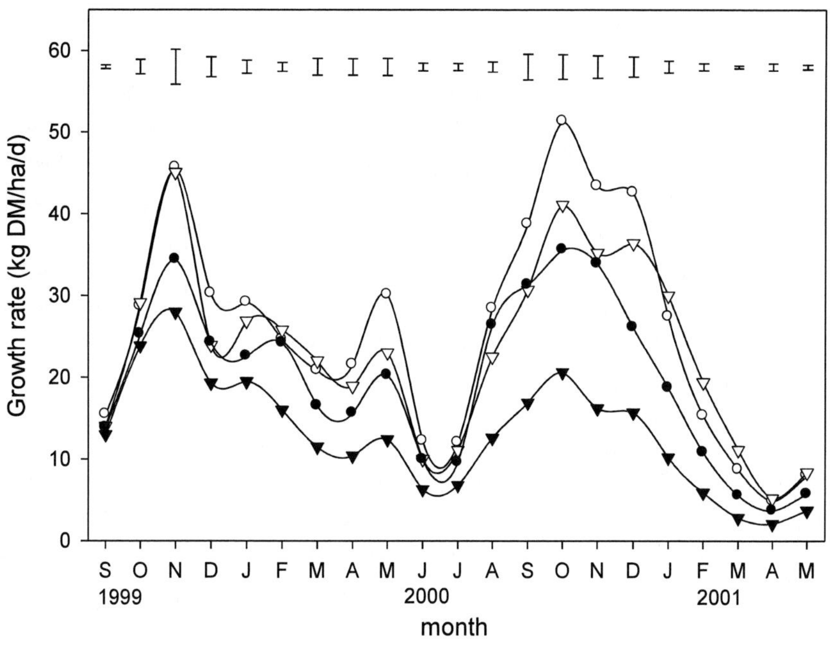

Figure 1. Orchardgrass dry matter growth rates (21±1 days regrowth) over time for four shade treatments at the Lincoln University silvopastoral experiment (Canterbury, New Zealand): open (○) (100% PPFD), open+slats (▽) (~43% transmissivity), under trees (●) (~58% transmissivity) and trees+slats (▼) (~24% transmissivity). Bars indicate standard error of the mean (sem) (Peri et al., 2007a).

For orchardgrass plants, the seasonal input of solar energy as photosynthetically active radiation (PAR) is the main determinant of growth when factors such as nutrients, water and temperature are non-limiting. Peri (2002) demonstrated that orchardgrass DM growth rate decreased by 20% under tree shade (~24% transmissivity) compared with the full sunlight pastures during periods of non-limiting water (irrigated pastures) and temperature. Thus, tree shade reduced total annual yield by ~3.2 t DM/ha/yr.

Estimations of annual radiation use efficiency (RUE) based on final DM yield and PPFD intercepted by canopy at the final harvest showed that orchardgrass plants grown in full sunlight was always lower than those estimated under different shade conditions. The mean RUE in full sunlight was 0.43 g DM MJ^{-1} PAR, whereas under shade varied from 0.55 to 0.81 g DM MJ^{-1} PAR under moderate (~58% transmissivity) and severe (~24% transmissivity) shades, respectively.

The extent of overstorey shading can alter the efficiency of energy conversion to DM by affecting light interception and the photosynthetic activity of individual leaves (Sheehy and Cooper, 1973). In field environments plants can experience frequent fluctuations in irradiance from full sun to shade caused by cloud cover, overstory shading (e.g. silvopastoral systems) and within canopy shading (Knapp and Smith, 1987). Therefore, to quantify changes in carbon gain (or DM production) of a canopy experiencing fluctuating light regimes, responses of photosynthetic activity of individual leaves under this regime must be understood.

The effects of different uniform light energy levels on leaf photosynthesis has been reported for orchardgrass. In controlled environment conditions, Frank and Barker (1976) reported an increment in the rate of net photosynthesis of about 80% from 200 to 1160 μmol m^{-2} s^{-1} PAR. Similarly, Eagles and Treharne (1969) reported that the photosynthetic rate on a chlorophyll basis was 60% higher as light intensity increased from 48 to 144 W m^{-2} for a natural Norwegian population of orchardgrass. Unfortunately, a continuous light regime does not reflect the fluctuating light, with periods of full sunlight and shade, that understorey plants experience in a silvopastoral system. The maximum light-saturated photosynthetic rate (Pmax) and stomatal conductance (gs) of field grown orchardgrass leaves in a silvopastoral system were measured at different times under moderate (850-950 μmol m^{-2} s^{-1} PPFD) and severe shade (85-95 μmol m^{-2} s^{-1} PPFD) (Peri et al., 2002b). Also Pmax and gs were measured after 30, 60 and 180 minutes of severe shade to determine the lag in the rise of photosynthesis rate from low to high irradiance levels (induction state). The highest Pmax and gs values obtained were 26.5 μmol CO_2 m^{-2} s^{-1} and 0.41 mol H_2O m^{-2} s^{-1} in non-limiting conditions with full sunlight (1900 μmol m^{-2} s^{-1} PPFD). The photosynthetic rate of individual orchardgrass leaves from high to low light intensities decreased as a function of the magnitude and duration of the PPFD level previously experienced (figure 2). The time required to reach full induction on return to full sun ($Pmax_s$= 1) was 15 minutes after 30 minutes of severe shade and 37 minutes after 180 minutes of shade (table 1). The rate of change of gs_s was slower than for $Pmax_s$ on entering shade and also slower during the subsequent induction process. This shows that factors other than stomatal closure cause the reduction in Pmax during the first five minutes under shade (Peri et al., 2002b). Consequently, differences in the dynamics of non-stomatal and stomatal responses to fluctuating light can determine the capacity of a particular species to utilise the incoming radiation in silvopastoral systems.

The rate of decrease in Pmax under different shade intensities and the responses during induction could be used as physiological indicators to define shade-tolerant species from a physiological perspective. Thus, species with a slow decrease in Pmax when exposed to shade, or fast responses to induction (higher values of IS_1 and IS_{10}), would be classified as more shade-tolerant because they would increase the carbon photosynthetic gain. It could be expected that species with faster stomatal opening during the slow phase of induction would allow greater sun utilization of fluctuating sun in silvopastoral systems. Orchardgrass showed a fast induction response being defined as a 'shade tolerant' species from a physiological perspective (Peri et al., 2002). However, under grazing regimes the concept of 'shade tolerance' may be more appropriate if the morphological response of the pasture to changes in light quality and quantity under tree shade, together with changes in feeding value, are considered. A more useful criterion to evaluate the shade tolerance of pastures under grazing regimes would be the ecological stability and persistence of a pasture together with animal production per hectare (Peri et al., 2001b).

Figure 3 shows the dynamic of the simulated maximum saturated gross leaf photosynthetic rate ($Pmax_G$) of orchardgrass leaves for two fluctuating light regimes (20 and 80% PPFD) (Peri et al., 2006). *Pmax* in the fluctuating light regime of 80% transmissivity (figure 3a) reached a maximum value (1.21 mg CO_2 m^{-2} s^{-1}) for 2.7 h during the full sunlight period and only declined to a minimum of 0.6 mg CO_2 m^{-2} s^{-1} (figure 3c). For the 20% transmissivity regime (figure 3b), *Pmax* reached the maximum value of 1.21 mg CO_2 m^{-2} s^{-1} for only 30 minutes of the day and declined to the lowest minimum of 0.44 mg CO_2 m^{-2} s^{-1} (figure 3d). The success of this approach for predicting Pmax is that it can be used in environments outside those in which the equations were derived (Peri et al., 2002b, 2006). Thus, by calculating the time course of shade for a particular tree canopy in a silvopastoral system it is possible to use these functions for situations of up to 180 minutes of shade.

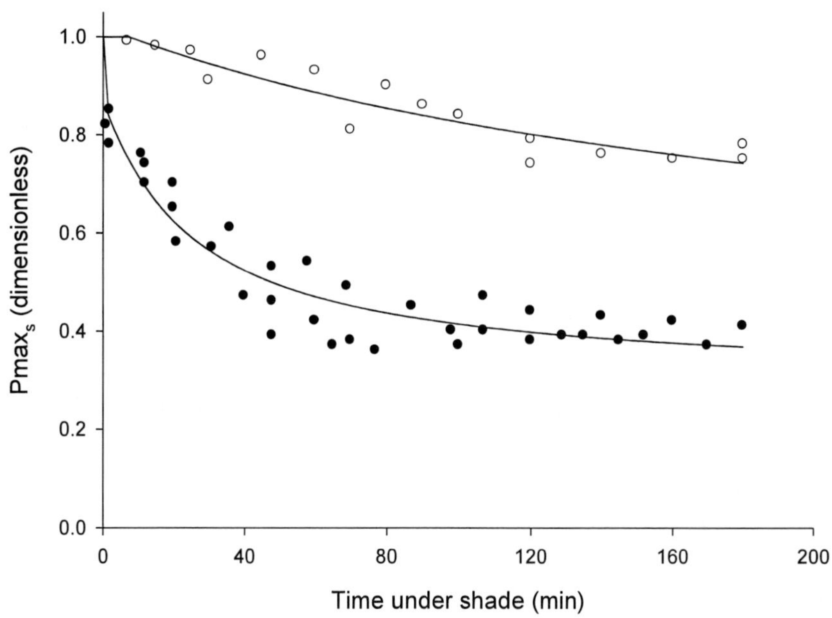

Figure 2. Time course of standardised rate of net photosynthesis ($Pmax_s$) for orchardgrass grown under field conditions in response to two light intensities: (•) 85-95 μmol m^{-2} s^{-1} PPFD and (○) 850-950 μmol m^{-2} s^{-1} PPFD. $Pmax_s$= 1 corresponds to Pmax= 26.5 μmol CO_2 m^{-2} s^{-1}. (Peri et al., 2002b).

Table 1. Induction state (IS) (%) at 1, 2 and 10 minutes from light increase (full sun at 1700-1900 μmol m^{-2} s^{-1} PPFD) for orchardgrass plants exposed to 30, 60 and 180 minutes of slat shade at 85-95 μmol m^{-2} s^{-1} PPFD (Peri et al., 2002b).

Time under shade (minutes)	IS_1	IS_2	IS_{10}
30	84	86	94
60	67	79	89
180	55	67	75
significance	***	***	***

*** $p<0.001$

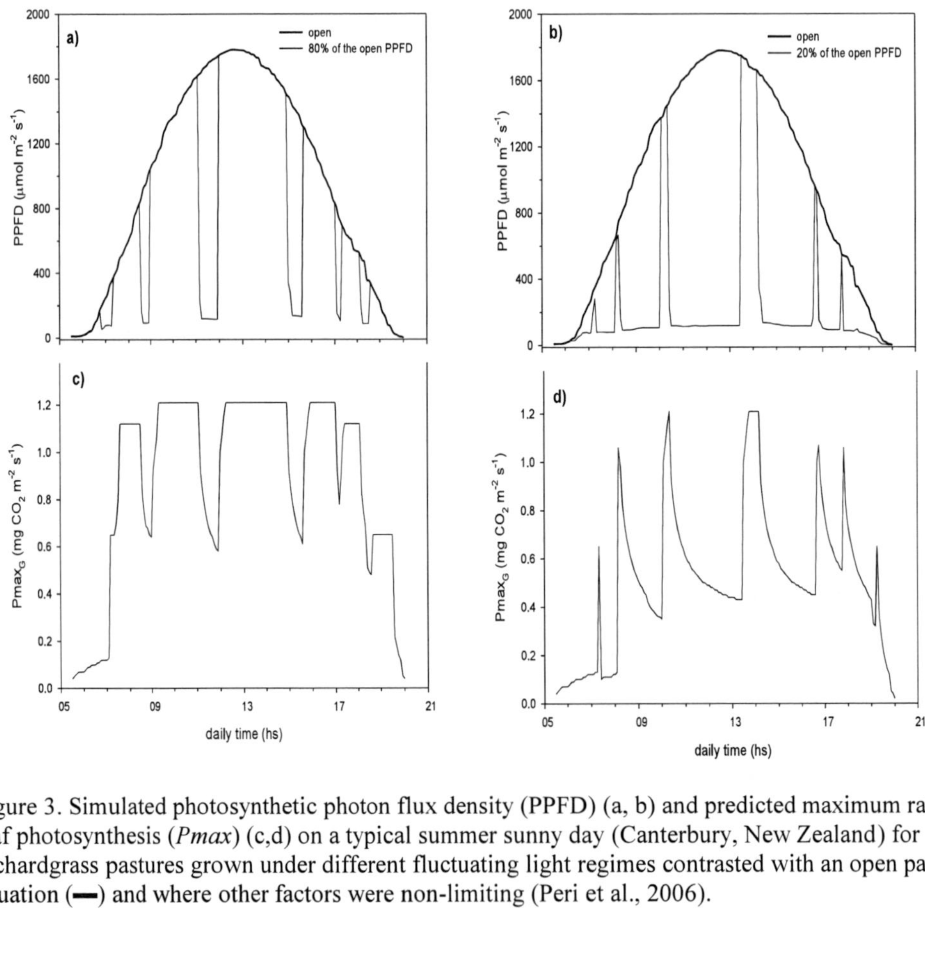

Figure 3. Simulated photosynthetic photon flux density (PPFD) (a, b) and predicted maximum rate of leaf photosynthesis (*Pmax*) (c,d) on a typical summer sunny day (Canterbury, New Zealand) for orchardgrass pastures grown under different fluctuating light regimes contrasted with an open pasture situation (**—**) and where other factors were non-limiting (Peri et al., 2006).

2.2. Nitrogen

Suboptimal nitrogen (N) is the most limiting nutrient for DM production in plants because it reduces productivity indirectly through restriction of leaf size and consequently canopy light interception and directly by reducing photosynthesis rates per unit leaf area

(Belanger, 1998). The response of a grass sward to fertiliser N applied at a range of different rates has been examined in numerous field trials. In a review, Whitehead (1995) reported that as the rate of applied N increased, there was an almost linear increase in herbage yield of 15-20 kg DM/kg N applied until application rates reached 250-400 kg N/ha/yr. High values of DM production for orchardgrass has been reported in several countries with the application of N fertiliser. In France, Lemaire et al. (1983) reported for orchardgrass a potential growth rate of 154 kg DM/ha/d with the application of 210 kg N/ha. In Finland, Rinne (1978) reported for orchardgrass pastures fertilised with 300 kg N/ha and grazed with dairy cows a mean DM production of 139±14 kg DM/ha/d during 34 days of regrowth.

A lack of clover and presence of obvious green urine patches in orchardgrass pastures can be used to indicate that they are nitrogen stressed. The impact of N on orchardgrass was reported by Peri et al. (2001b) who showed a mean maximum growth rate of 130 kg DM/ha/d in October-November for orchardgrass new urine patches (N applied from urination varied from 173 in autumn to 448 kg N/ha) in spring which was three times higher than the non-urine areas, but these differences decreased in summer and autumn (figure 4). Joshi et al. (1999) who showed a 42% increase in orchardgrass DM production under both moderate shade (67% of the open PPFD) and full sunlight in urine patches compared with adjacent non-urine patches.

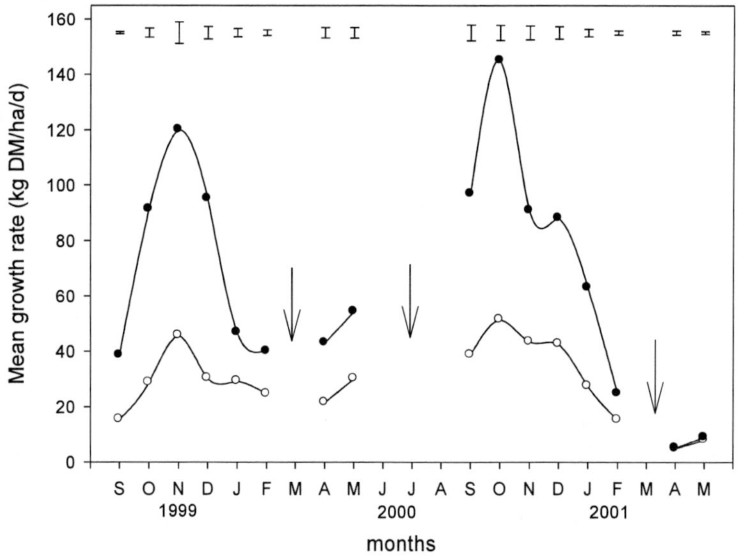

Figure 4. Mean orchardgrass dry matter growth rate after 21±1 days regrowth from new urine (●) and paired non-urine (○) patches. Arrows indicate discontinuity in grazing. Bars indicate standard error of the mean (sem) (Peri et al., 2001b).

Nitrogen use efficiency (NUE) (kg DM/kg N) is commonly used to determine agricultural efficiency. The NUE, based on N recovered and including the contribution of soil mineral N, for orchardgrass grown in the open (100% transmissivity) and in a silvopastoral system under 10 year old Pinus radiata tree shade (~58% transmissivity) at 200 stems/ha (Lincoln University silvopastoral experimental area, Canterbury, New Zealand) is shown in table 2. The results showed that NUE for orchardgrass pasture varied from 19.2 to 46.9 kg

DM/kg N depending on the light intensity, water availability, N fertilizer and regrowth period. This is consistent with Smith et al. (2000) who highlighted that this approach is environment specific, highly variable and dependent on the time of application and form of N fertilizer used. Thus, extrapolation of these results may induce excess of N applications for orchardgrass requirements due to the differences in the main environmental drivers such as temperature, rainfall and/or solar radiation between sites, seasons and years.

Quantification of expected DM responses to applied N may be improved using a nitrogen dilution curve ($N(\%) = a(DM)^{-b}$) that accounts for the decrease in herbage N concentration (%N) as yield accumulates (Lemaire, 1997). This shows the reduction in herbage N% as DM yield (t DM/ha) accumulates for a vegetative crop supplied with non-limiting N and sufficient water for growth. Mills et al. (2009) reported a N dilution curve ($N\% = 4.8DM^{-0.35}$) that indicated that for a 1.0 tDM/ha yield of green orchardgrass herbage the optimum N concentration (N_{opt}) was 4.8%, but at 3.0 tDM/ha the N_{opt} was 3.3%, or 99 kg N/ha in green herbage.

Table 2. Nitrogen use efficiency (NUE) (kg DM/kg N) for different regrowth periods and annual for orchardgrass pastures at two light (open and tree shade of ~58% transmissivity), two irrigation (0 or fully) and two nitrogen (0 or 300 kg N/ha) levels. Regrowth periods were 60 days for spring, summer and autumn and 110 days for winter (May-August), Lincoln University silvopastoral experimental area in Canterbury, New Zealand.

Treatment	Sep-Oct	Nov-Dec	Jan-Feb	Mar-Apr	May-Aug	Mean annual NUE
Open						
control	29.0	42.8	46.9	35.3	26.0	36.0
Open W	29.0	39.8	43.1	30.3	26.0	33.6
Open N	21.0	25.5	27.9	19.9	19.6	22.8
Open W+N	21.0	26.2	27.3	20.6	19.6	22.9
Trees						
control	28.9	42.6	46.7	20.5	26.9	33.1
Trees W	28.9	37.9	40.8	29.2	25.5	32.5
Trees N	21.3	25.1	26.2	20.0	19.7	22.5
Trees W+N	21.3	25.2	26.0	20.0	19.2	22.3

A positive linear or curvilinear relationship between leaf N% and Pmax has been reported for several species (Field, 1983; Hirose and Werger, 1987; Hilbert et al., 1991). Specifically, Peri et al. (2002a) reported the effect of leaf N% (from 1.5 to 5.9%) on the rate of net photosynthesis (Pmax) in irrigated orchardgrass plants (no water stress) within the defined optimum temperature range (19-23 °C). The same maximum value of 27.4 μmol $CO_2\ m^{-2}\ s^{-1}$ ($Pmax_s$= 1) was measured from 5.2 to 5.9% N. From this point *Pmax* decreased at an average rate of 3.0 μmol $CO_2\ m^{-2}\ s^{-1}$ per 1% N down to 2.6% leaf N content (figure 5). This was followed by a further decline of 11.3 μmol CO2 m-2 s-1 per 1% N down to the lowest recorded value of N%. The lack of relationship between Pmax and gs indicates factors other than stomatal conductance caused the reduction in net photosynthesis. The generality of the N-leaf photosynthesis relationship strongly suggests that one or several nitrogenous leaf components directly limit photosynthetic capacity. The effect of N on Pmax per unit leaf area can be explained by the increment of chloroplast content. For example, the application of 300 kg N/ha increased the total chlorophyll content in the orchardgrass leaves from 0.55 g/m^2 in irrigated plants with N deficiency to 0.96 g/m^2 in non-N deficiency plants for the 21 day regrowth period (Peri, 2002). This can be interpreted as giving a greater capacity for light absorption and increasing Pmax per unit leaf area. It follows that decreased chlorophyll formation during nitrogen deficiency decreased the photosynthetic potential of the leaf by decrease protein content and activity (Evans, 1996).

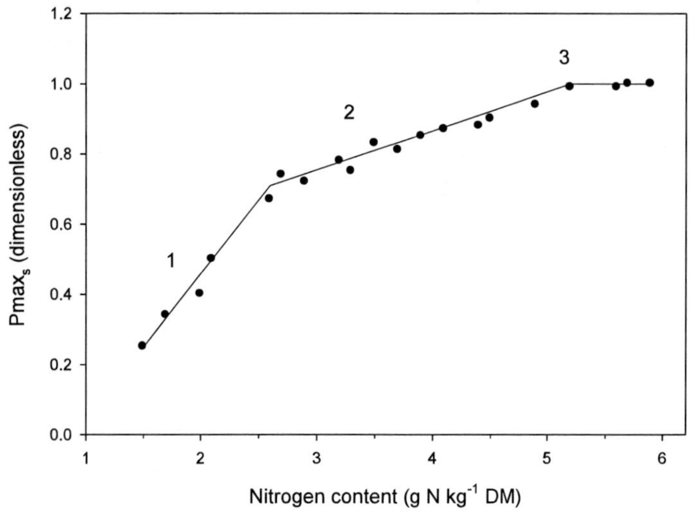

Figure 5. Response of the standardised rate of net photosynthesis (Pmax$_s$) to nitrogen content for orchardgrass grown under field conditions where temperature and water were non-limiting. *Pmax$_s$*= 1 corresponds *Pmax*= 27 μmol CO_2 m^{-2} s^{-1}. Fitted three part (1, 2 and 3) "broken stick" model (—) is indicated (Peri et al., 2002a).

2.3. Water

Orchardgrass is a widespread perennial grass, which is well-adapted to dry conditions and can survive a soil water deficit more effectively than most temperate forage grasses (Volaire and Thomas, 1995). Stevens et al. (1992) reported that under a lax grazing system in a dry Canterbury site, 'Kara' orchardgrass produced 55% more DM than 'Nui' ryegrass during summer. Similarly, Lancashire and Brock (1983) reported that 'Grasslands Wana' orchardgrass pasture produced 62 kg DM/ha/d compared with 42 kg DM/ha/d from 'Nui' ryegrass during summer on a dry hill country site. This is assumed to be because of the deep rooting system of orchardgrass which can withdraw moisture from a greater soil depth and also for the more effective dehydration delay associated with slower decrease in photosynthesis activity and improved carbon balance (Volaire and Thomas, 1995).

On other hand, orchardgrass is also responsive to irrigation. Volaire and Thomas (1995) reported that irrigated orchardgrass plants (watered to field capacity every 2 days) produced 3 times more aerial biomass than stressed plants (80 days droughted plants with a leaf water potential of –30 bar) after 60 days regrowth. Penman (1962) reported that irrigated 'S37' orchardgrass produced a mean of 25% more DM than controls (soil moisture deficit > 50 mm) over six years evaluation at Woburn (England). Peri (2002) reported that, on average, the application of irrigation increased the annual yield of orchardgrass pastures by ~4.4 t DM/ha/yr (Canterbyry, New Zealand). The growth rate was 15 kg DM/ha/d after day 60 of the January-February regrowth period when maximum water stress occurred (soil VWC of 14% in unirrigated plots) compared with 38 kg DM/ha/d for irrigated pastures.

From data reported by Peri (2002), the water use efficiency (kg DM/mm) was determined from the slope of linear regression between accumulated annual DM yield and cumulative water use (mm) from 0 to 0.5 m soil depth for orchardgrass growing at two light (open and tree shade of ~58% transmissivity), two irrigation (0 or fully) and two nitrogen (0 or 300 kg N/ha) levels (Canterbury, New Zealand). Orchardgrass WUE ranged from 12.1 kg DM/mm for pasture grown under the tree shade without irrigation and fertilizer (Trees control) to 38.9 kg DM/mm for pastures growing with N fertilizer in the open (table 3). There was also an increased of WUE in response to water stress (table 3). Despite this variation, N was the main factor affecting orchardgrass WUE. This is consistent with Mills et al. (2009) who reported from a comparison of N uptake ratio with WUE from orchardgrass pastures that N deficient pastures used available water less efficiently in the production of DM than pastures supplied with non limiting N. With increasing soil moisture deficits the N uptake ratio decreased which consequently reduced WUE.

Table 3. Mean annual water use efficiency (WUE) (kg DM/mm) for orchardgrass growing at two light (open and tree shade of ~58% transmissivity), two irrigation (0 or fully) and two nitrogen (0 or 300 kg N/ha) levels. Canterbury, New Zealand.

Treatment	Mean annual WUE (kg DM/mm)
Open control	18.2 ±0.25
Open W	16.6 ±0.38
Open N	38.9 ±0.55
Open W+N	32.1 ±0.41
Trees control	14.1 ±0.17
Trees W	12.9 ±0.22
Trees N	31.5 ±0.49
Trees W+N	25.0 ±0.22

The physiological basis for differences between species in WUE is a result of the net photosynthetic ability of the plant under non-limiting conditions. In general, water stress has a negative effect on leaf photosynthesis. Peri et al. (2002a) reported the effect of water stress (expressed as pre-dawn leaf water potential (ψ_{lp})) on net photosynthesis (Pmax) of orchardgrass leaves grown non limiting water and nitrogen conditions. There was a strong negative relationship between Pmax and ψ_{lp} (figure 6). The maximum value of Pmax of 27.4 μmol CO_2 m^{-2} s^{-1} ($Pmax_s$= 1) was measured from –0.1 to –1.2 bar (from 30 to 27% soil VWC). From this point, Pmax decreased linearly at the rate of 2.1 μmol CO_2 m^{-2} s^{-1} per bar of ψ_{lp} as water stress increased down to –14.0 bar. Then, Pmax reached a constant negative value (from -0.1 to -0.5 μmol CO_2 m^{-2} s^{-1}) which indicated that total respiration was higher than photosynthesis under severe water stress. Effects of water stress on net photosynthesis can be caused by stomatal and non-stomatal factors. Peri et al., 2002a reported that the linear reduction in stomatal conductance to water vapor was the main factor that reduced Pmax.

However, it is now recognized that stomata do not respond to changes in leaf water potential until a critical threshold level is reached. Jackson (1974) reported that a field value for ψlp of –15.0 bar gave about a 70% decrease in leaf stomatal diffusion rate for orchardgrass plants.

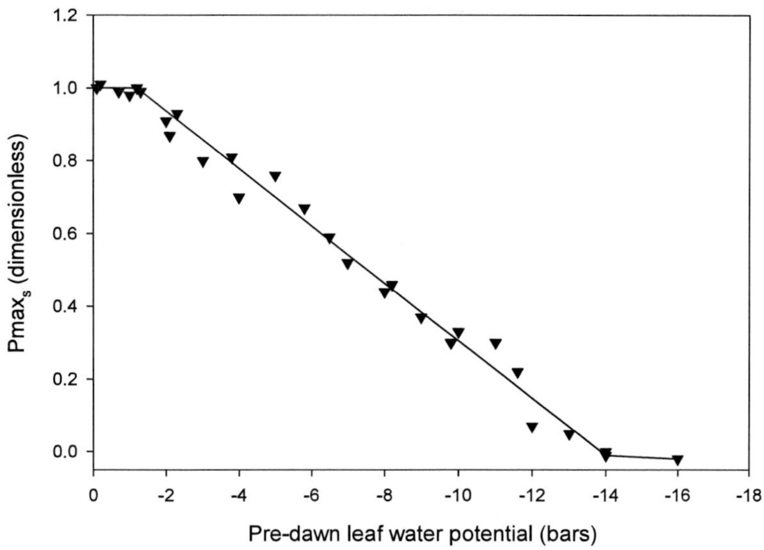

Figure 6. Response of the standardised rate of net photosynthesis ($Pmax_s$) to water stress for orchardgrass grown under field conditions where temperature and N were non-limiting. $Pmax_s$= 1 corresponds $Pmax$= 27 μmol CO_2 m^{-2} s^{-1} (Peri et al., 2002a).

2.4. Temperature

Temperatures below and above the optimum for a plant affect phenological, morphological and physiological processes and therefore its DM production.

Mills et al., (2009) reported linear relationships between DM accumulation against accumulated thermal time (°Cd) with a base air temperature (Tb) of 3 °C for a 'Wana' orchardgrass monoculture at Lincoln University, Canterbury New Zealand. They found that N fertilized pastures produced DM at a near constant rate of 7.0 kg DM/◦Cd/ha above Tb when periods of water stress were excluded. For no fertilized pastures, the rate was 56% lower at 3.3 kg DM/◦Cd/ha. However, in silvopastoral systems the decrease in DM production with shade intensity showed seasonal variation responses with less difference during winter (mean daily air temperatures < 8 °C) indicating that pasture production during winter was limited mainly by low temperatures (Peri et al., 2007a).

The optimum temperature for growth of most temperate grasses is 20 to 25 °C (Robson et al., 1988). Eagles (1967) reported that the optimum range for orchardgrass leaf photosynthesis in controlled environments was 20-22 °C. Also under a controlled environment, Mitchell and Lucanus (1960) reported that orchardgrass growth at 7 °C during the day decreased by 78% compared with 15.5 °C. Knievel and Smith (1973) showed that temperatures above 28 °C greatly reduce orchardgrass growth. Peri et al. (2002a) reported that Pmax per unit of orchardgrass leaf increased by an average of 1.6 μmol CO_2 m^{-2} s^{-1} per °C from 10 to 19 °C and then plateaued at a Pmax of 27.4 μmol CO_2 m^{-2} s^{-1} from 19 to 23 °C. Pmax then declined by an average of 2.1 μmol CO_2 m^{-2} s^{-1} per °C from 23 to 31 °C (figure 7).

Extrapolation of the ends of the fitted "broken stick" function predicted that Pmax= 0 at 2 and 37 °C. The poor relationship between Pmax and gs (Peri et al., 2002a) suggests that a change in stomatal conductance was not responsible for the reduction in Pmax. At low temperatures the reduction in Pmax cannot be accounted for by stomatal limitations (Nie et al., 1992). Thus, low temperature-induced inhibition probably reflects changes at the chloroplast level rather than limitations on leaf gas exchange. At high temperatures (above 23 °C), it is likely that the photorespiration rate increases with temperature faster than net photosynthesis (Peri et al., 2002a). This is consistent with Knievel and Smith (1973) who showed that temperatures above 28 °C greatly reduce orchadgrass growth.

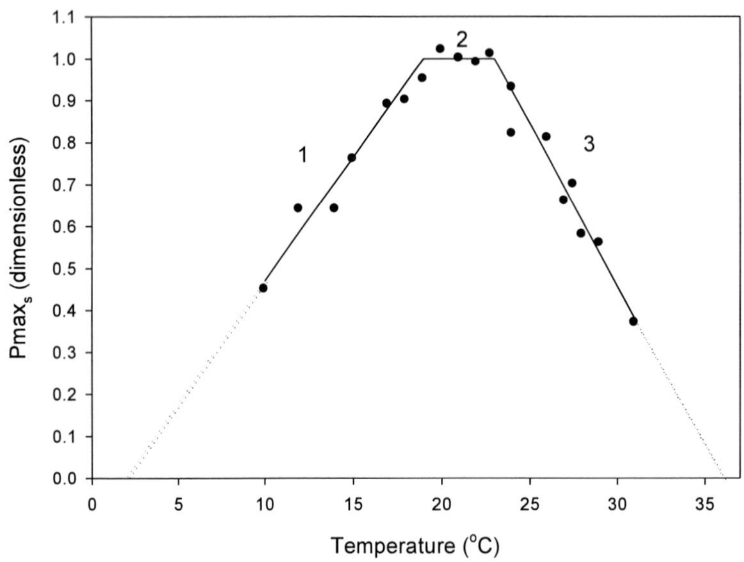

Figure 7. Response of the standardised rate of net photosynthesis (*Pmax$_s$*) to temperature for orchardgrass grown under field conditions where N and water were non-limiting. *Pmax$_s$*= 1 corresponds with *Pmax*= 27 μmol CO_2 m^{-2} s^{-1}. The fitted three part (1, 2 and 3) "broken stick" model (—) is indicated, with extrapolation to *Pmax$_s$*= 0 (····) (Peri et al., 2002a).

In silvopastoral systems, trees modify the air temperature and therefore alter the potential DM production of the understorey species. To determine which temperature to use for prediction of Pmax for orchardgrass pastures under shade conditions, air and canopy temperatures were compared (Peri et al., 2003b). Air temperature was higher than canopy temperature under full sun and tree shade conditions at midday, and this difference increased with increasing air temperatures. The magnitude of the difference between air and canopy temperature (T_{a-c}) varied between sun and shade situations according to exponential functions (Peri et al., 2003b). Between 10 and 20 °C air temperature, the mean T_{a-c} was 0.3 °C in sun plants compared with 2.6 °C under shade conditions. However, at an air temperature of 31 °C, the mean maximum value of T_{a-c} was 2.0 °C for plants in full sun and 7.4 °C for orchardgrass plants under shade. For this reason, canopy temperature was used to validate the multiplicative model of Pmax for orchardgrass plants under shade when air temperature was limiting (temperatures >23 °C).

2.5. Regrowth Duration

The productivity of a pasture is also dependent on management factors that affect the growth of the sward. Regrowth duration is a management factor that can be modified through the frequency and severity of defoliation (e.g. infrequent cutting for hay or silage, rotational or continuous grazing). In each growth period following cutting there is an initial lag phase, succeeded by a period of nearly constant linear growth and finally an asymptotic phase where leaf area exceeds optimal values (when 95% of light is intercepted) for the particular species (Brougham, 1956). Pearce et al. (1965) reported that, from irrigated and fertilized orchardgrass sward, the maximum growth was at LAI= 5.5 after approximately 20 days regrowth and then declined by 35% at an LAI of about 8.5. Peri (2002) showed that the DM growth rate of irrigated and N fertilized orchargrass pastures during the January-February regrowth period (mean air temperature 15 ºC), increased from 65 kg DM/ha/d at day 10 to the maximum 137 kg DM/ha/d at day 30 (LAI= 6.3), and then declined to 99 kg DM/ha/d after 60 days regrowth. The time at which maximum DM production occurred during regrowth periods, which corresponded approximately to a 95% DM accumulation, depended on seasons and on light intensity (Peri, 2002). Thus, maximum DM production occurred earlier with increments in air temperature and later under tree shade. As LAI increased so did light interception, causing increases in DM up to a critical LAI value of 6.0.

Peri et al. (2003a) reported that the rate of net photosynthesis of orchardgrass leaves as a function of PPFD showed an asymptotic response with P_{max} at 27.4 μmol $CO_2\,m^{-2}\,s^{-1}$ in non-limiting conditions (20 days regrowth) with leaf saturation at 1000 μmol $m^{-2}\,s^{-1}$ PPFD (figure 8A). However, P_{max} decreased with regrowth-duration and was 23.7 and 14.2 μmol $CO_2\,m^{-2}\,s^{-1}$ for successive youngest fully expanded leaves of orchardgrass plants grown for 40 and 60 days, respectively. From 20 to 25 days regrowth, P_{max} per unit of leaf was constant and maximal (P_{maxs}= 1) but it then decreased by 0.42 μmol $CO_2\,m^{-2}\,s^{-1}$ per day of regrowth (figure 8B). To provide a more mechanistic relationship, detailed analysis of shading within the canopy and tiller morphology may be required. For example, N and irrigation initially increase net leaf photosynthesis (Peri et al. 2002a). However, this stimulation of leaf expansion would result in mutual shading that may depress leaf photosynthesis as the regrowth period is extended. This is because leaves develop from the apex, which remains near the soil surface in vegetative swards. This is progressively shaded as LAI increases (figure 9A) so that successive leaves develop in increasingly shaded environments with a lower photosynthetic capacity. As a consequence, when they emerge into full sunlight they are unable to fully utilize the available light conditions (Woledge and Leafe, 1976; Sheehy, 1977). A second consideration is that successive youngest expanded leaves may actually be chronologically older as regrowth days progress. This is consistent with the measured tiller morphology over 60 days of regrowth, where lamina length and pseudo-stem height increased (figure 9B). The implication is that the youngest expanded leaf at day 60 was older than those at day 10. Duru and Ducrocq (2000) reported that as orchardgrass plants grew up to 80 days of regrowth (N and temperature non-limiting conditions), the leaf appearance rate per tiller decreased and the lamina growth duration, lamina length (from 14 to 44 cm) and life-span (from 362 to 580°Cd) increased with its insertion level (sheath length). Furthermore, because leaf appearance rate decreases and life-span increases during regrowth, the number of living leaves is fairly constant (Duru and Ducrocq, 2000). This was shown for orchardgrass with an almost constant 3.5 green leaves per tiller (figure 9B). This confirms the greater longevity of leaves at the end of a rotation. Thus, a dominant tiller at 60 days of age may be proceeding through a senescence process even though it may carry the youngest fully expanded leaf of the plant.

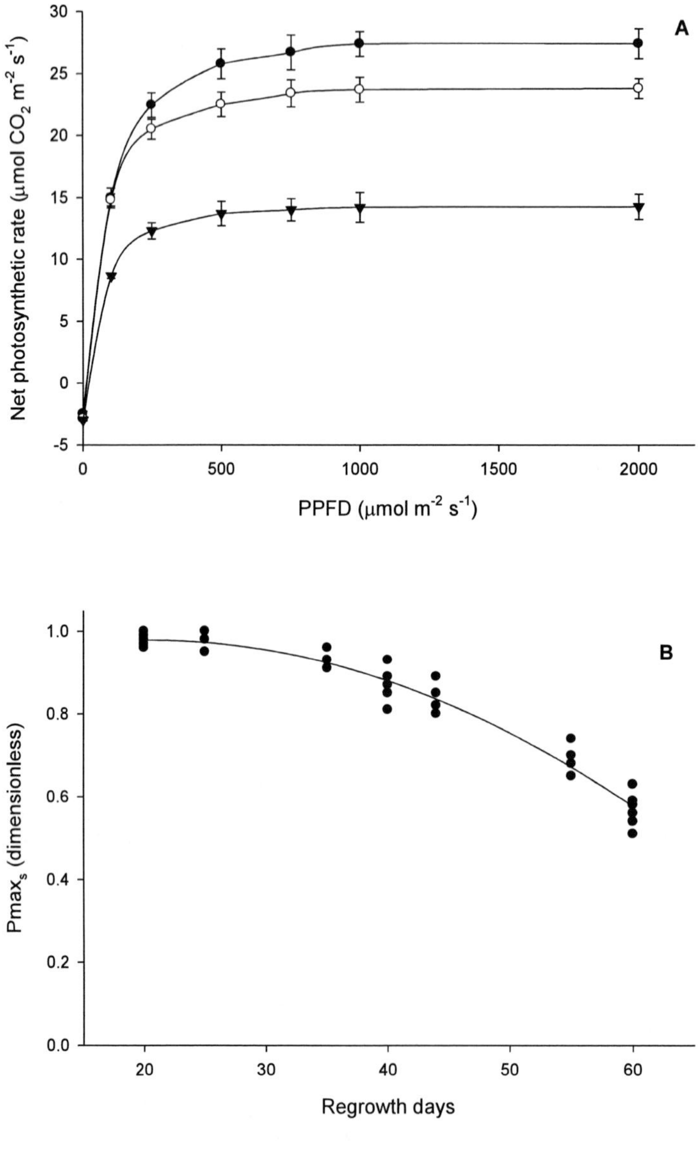

Figure 8. (A) Net photosynthetic rate against photosynthetic photon flux density (PPFD) for the youngest fully expanded leaf of orchardgrass grown in a field environment with regrowth duration non-limiting at 20 days (•) and then after 40 (○) and 60 (▼) days. Bars indicate standard error of mean (s.e.m.); (B) Standardised rate of net photosynthesis (P_{maxs}) against regrowth duration for orchardgrass grown under field conditions where temperature, nitrogen and water were non-limiting. P_{maxs}= 1 ≡ P_{max}= 27.4 μmol CO_2 m^{-2} s^{-1} (Peri et al., 2003a).

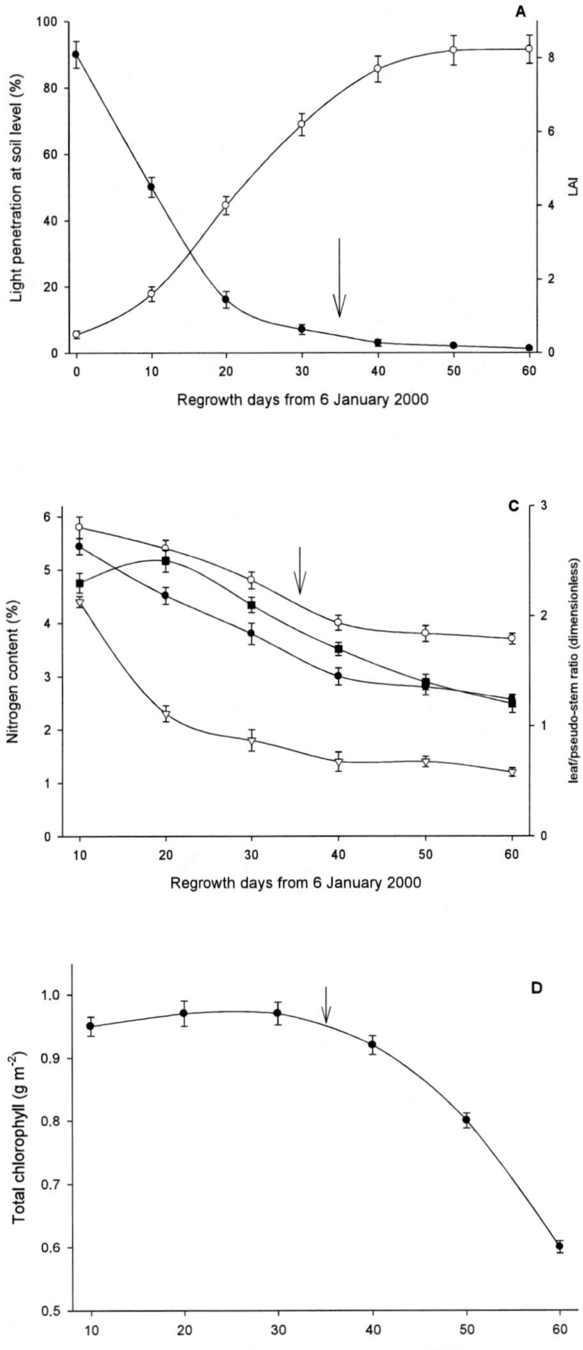

Figure 9. (A) Light penetration at soil level (●) and leaf area index (LAI) (○); (B) length of the youngest expanded leaf (○), pseudo-stem height (●) and number of green leaves per tiller (▼) of dominant tillers; (C) total herbage nitrogen content (●), leaf N content (○), pseudo-stem N content (□) and leaf:pseudo-stem ratio (■); and (D) mean total leaf chlorophyll content per unit of area (g m^{-2}) of orchardgrass pastures. All measurements were taken over a 60 day regrowth period in non-limiting temperature, nitrogen and water conditions. Bars indicate standard error of mean (s.e.m.). Arrows indicate the start of lodging (Peri et al., 2003a).

For grasses, the negative effect of leaf age on leaf photosynthesis in different positions on one tiller, and during ageing of leaves in a particular position on the tiller, has previously been reported (e.g. Treharne and Eagles, 1970; Woledge, 1972; Robson and Parsons, 1978). A likely cause of the decrease in P_{max} in the sward was decline in herbage and leaf N content over time (figure 9C). Even with an optimal N supply the plant N concentration declines with age or DM accumulation. This is because, in older plants, a greater proportion of resources is diverted to structural support and other non-photosynthetic material of low N content (van Keulen et al., 1989). This effect was highlighted by the increase in the relative amount of pseudo-stem compared with leaf DM or reduced leaf:stem ratio over time (figure 9C). Mobile nutrients, including N, are partially remobilized from senescing leaves and translocated to other plant parts during the ageing process (Whitehead, 1995). The reasons for a strong relationship between P_{max} and leaf N content were reported previously (Peri et al., 2002a). These were mainly the increase in soluble proteins (predominantly enzymes) and the increase of compounds associated with the light reactions (including chlorophyll). The decrease in leaf N content with regrowth was consistent with the decrease in chlorophyll content of the youngest expanded leaf (figure 9D). Treharne et al. (1968) also found that leaf chlorophyll content in orchardgrass decrease with age. Treharne and Eagles (1970) reported a fall of 60% in RuBisCO activity of the youngest fully expanded orchardgrass leaves after 30 days full expansion at 25ºC. Another reason for the decline in P_{max} as leaves age, or as regrowth-duration progresses, is the decrease in gs. Peri et al. (2003a) found that gs decreased linearly by 30% from 20 to 60 days regrowth and support the hypothesis that the youngest expanded leaf at day 60 was older than that at day 20.

3. Morphology and Leaf Area Index (LAI)

Pasture growth also varies according to total canopy leaf area index (LAI) and the arrangement of the angular distribution of leaves (i.e. the canopy architecture). Together these determine the interception of solar radiation by a pasture and the distribution of irradiance among individual leaves (Sheehy and Cooper, 1973; Peri et al., 2005b, 2006). Maximum pasture production requires complete capture of incident solar radiation and can only be achieved with supporting levels of water and nutrients and non-limiting temperatures. LAI, which depends on the rate of leaf appearance, growth and death of individual tillers and leaves and their morphological changes, has been reported to be dependent on temperature, irradiance, water status (Davies, 1988) and light quality (Casal et al., 1987; Peri et al., 2007a).

As for DM production, Peri et al. (2007a) reported that orchardgrass LAI curves showed seasonal fluctuations (figure 10A). The greatest LAI occurred in spring during October-November and there was a rapid decrease in late summer and winter. The LAI was consistently lower under trees (~58% transmissivity) and severe shade (~24% transmissivity) compared with the full sunlight conditions in all seasons. Change in light quantity and quality under trees can modify LAI because stem elongation can be promoted and tillering and branching inhibited (Garnier and Roy, 1988). Peri et al. (2007a) showed that shade encouraged vegetative orchardgrass plants to become etiolated (figure 10B) where the taller growth may be an effort to gain greater access to available light in competition with neighbouring plants and tillers. For example, when water was non-limiting (September-

November), shade increased canopy height by approximately 23% for pastures grown at ~43-58% transmissivity and by 41% for ~24% transmissivity conditions (figure 10B). It also appears that the etiolation of shaded orchardgrass plants, responded to a reduction in the R:FR ratio. Peri et al. (2007a) reported a differential absorption of red and far-red light from the tree canopies produced a 56% decrease in the R:FR ratio in the middle of the tree shade compared with full sunlight. The changes in R:FR ratio are perceived by understorey plants through the phytocrome system which may change morphogenetic characters in plants (Smith, 1982). It has been demostrated that leaf area of shaded orchardgrass leaf blades is increased to maximise light interception at the expense of leaf thickness, resulting in leaves being longer, narrower, and thinner than when grown in full sunlight conditions (Peri et al., 2005a). This is consistent with Devkota et al. (2000) who reported that plants from 10 orchardgrass selections increased the specific leaf area with shade from 15.9 $mm^2 mg^{-1}$ under 73% of the open PPFD to 21.3 $mm^2 mg^{-1}$ under 24% of the open PPFD.

Reduced light intensity and changes in light quality have been reported to reduce tillering and are therefore likely to reduce LAI. For example, orchardgrass vegetative tiller population decreased from 5540 in the full sunlight to 3570 tillers m^{-2} under severe shade (~24% transmissivity) (figure 10C). Similarly, Garnier and Roy (1988) reported a 36% reduction of orchardgrass tiller population in France under 33% transmissivity oak tree shade compared with open pasture. Devkota et al. (1997) reported for a range of orchardgrass cultivars that the mean tiller number declined 25-30% as the shade environment fell from 77 to 17% PPFD of full sunlight. The physiological basis for the reduction in tillering is that under low irradiance a reduced supply of current assimilate is preferentially allocated to existing tillers at the expense of axillary buds (Robson et al., 1988). Thus, the effect of low light intensity is not on the rate of site production, but rather on the extent to which sites are filled.

Also, the addition of N and irrigation had a positive effect on LAI in open and shaded orchardgrass pastures (Peri, 2002). For example, after 60 days of regrowth during the January-February period (temperature no limiting), LAI ranged from 8.2 (in irrigated and N fertilised open pastures) to 2.5 (control pastures under trees) (figure 11A).

The irrigation had a positive effect on LAI in open and shaded pastures increasing the radiation interception of a sward by through a greater leaf expansion and therefore pasture height (figure 11B) and enhancing tillering (figure 11C). This is consistent with Hsiao and Acevedo (1974) who reported that the cell expansion is sensitive to water stress and with Lawlor (1972) who showed an 80% reduction in leaf elongation rate of perennial ryegrass when leaf water potential fell from –4 to –10 bars, and elongation ceased at –16 bars. Norris (1982) found for three moisture treatments and a range of grasses that irrigation (maximum potential soil moisture deficit, MSMD, of 41 mm) increased tiller number over control (MSMD of 239 mm) and covered (MSMD of 273 mm) plots. Irrigated plots had higher tillering rates (0.037 tillers $tiller^{-1} d^{-1}$) than covered plots (0.010 tillers $tiller^{-1} d^{-1}$), while control plots were intermediate (0.018 tillers $tiller^{-1} d^{-1}$).

The influence of N supply on pasture growth has been reported to increase LAI of the sward through an increase in the rate of leaf extension (Wilman and Wright, 1983). The fast increment of canopy development (LAI) due to N can also be explained by increases in tiller population (figure 11C) and canopy height (figure 11B). Auda et al. (1966) showed that the number of tillers of orchardgrass grown in soil/sand mixtures was three times greater when 224 kg N/ha was applied than without an application of N.

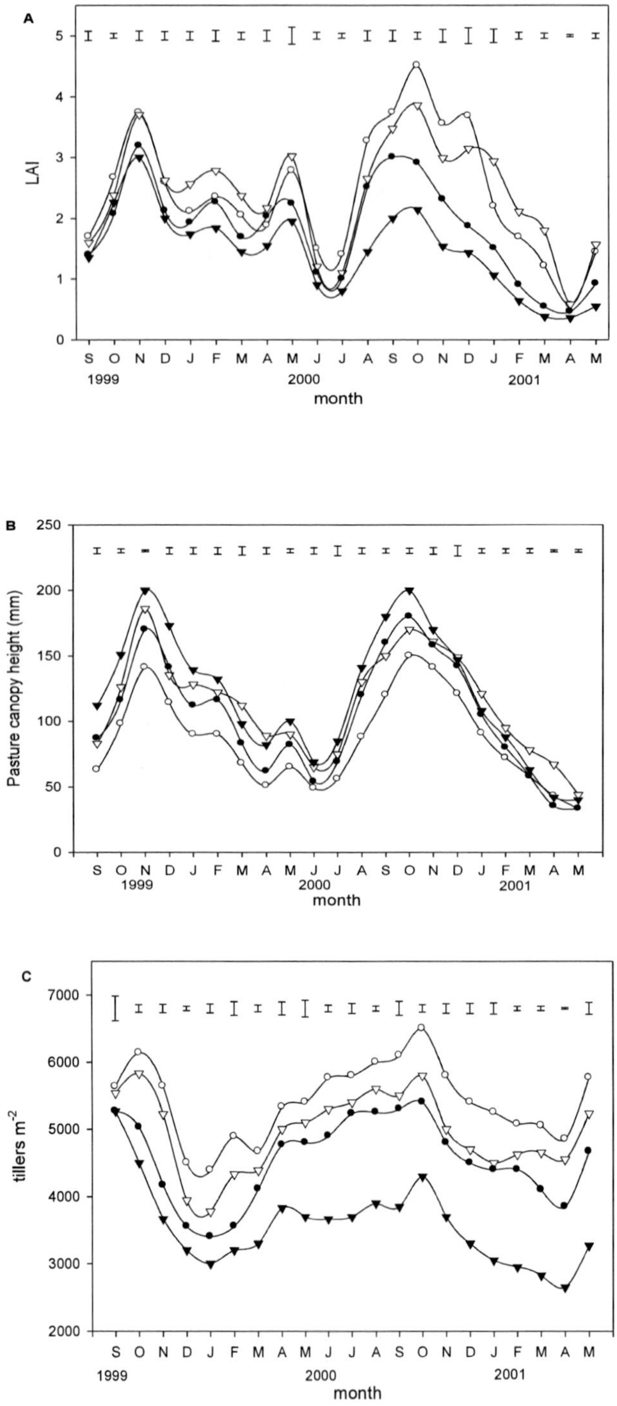

Figure 10. Orchardgrass leaf area index (LAI) (A), canopy height (B) and tiller population (C) (21±1 days regrowth) over time for four shade levels at Lincoln University silvopastoral trial (Canterbury, New Zealand): open (○) (100% transmissivity), ~43% transmissivity (▽), ~58% transmissivity (●) and ~24% transmissivity (▼). Bars indicate standard error of the mean (Peri et al., 2007a).

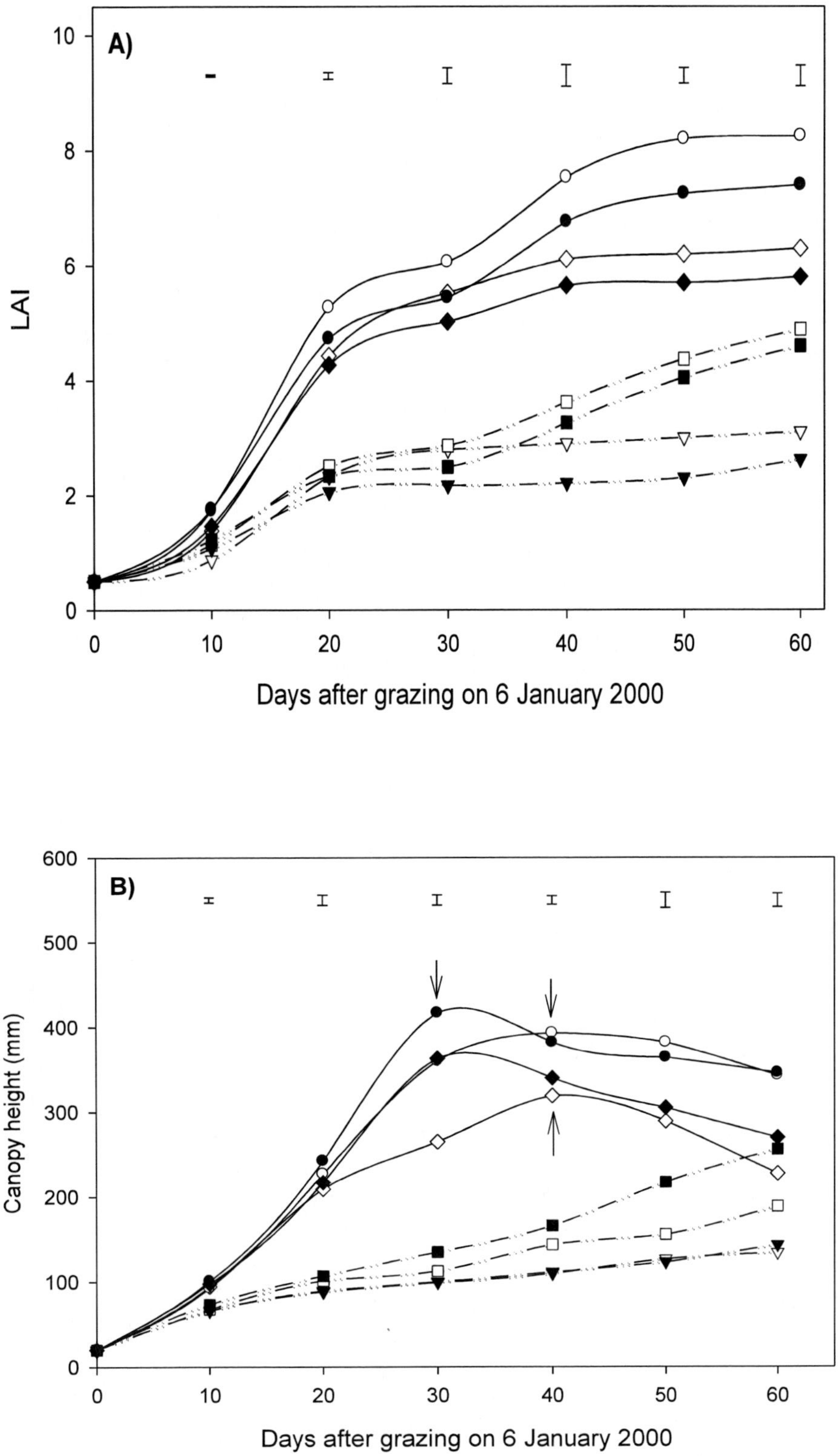

Figure 11. Continued).

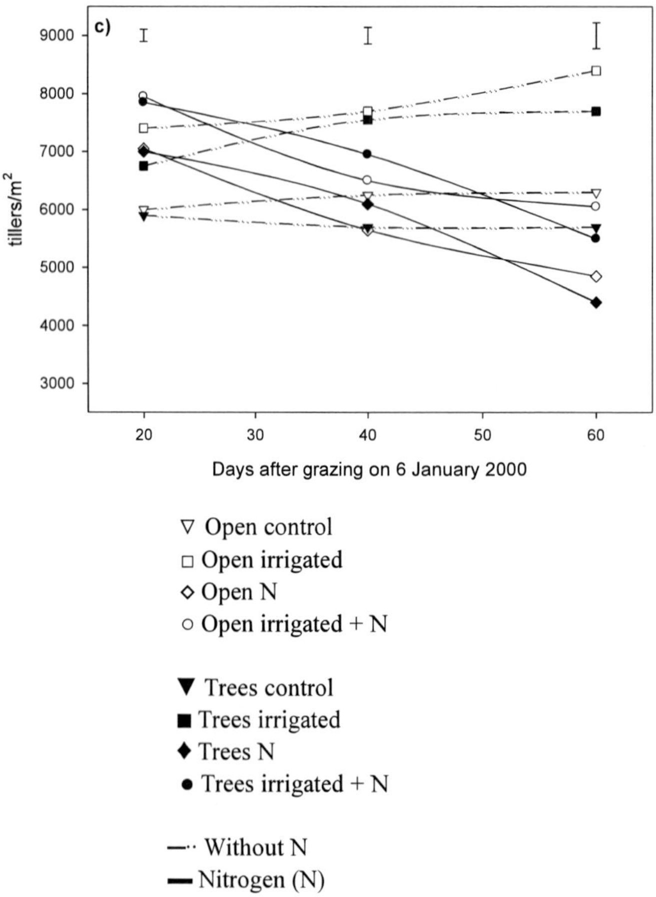

Figure 11. (A) leaf area index (LAI); (B) canopy height; (C) tiller population of orchardgrass pastures grown at for two levels of light intensity (open pasture: 100% transmittance or pasture under tree shade: ~58% transmissivity), two levels of irrigation (0 or fully) and two levels of nitrogen (0 or 300 kg N/ha). All measurements were taken over a 60 day regrowth period in non-limiting temperature, nitrogen and water conditions. Bars indicate standard error of mean (s.e.m.). Arrows indicate the start of lodging (Peri, 2002).

There is a non-linear relationship between DM yield and LAI (from 0.5 to 8.3) for the vegetative orchardgrass pasture (Peri, 2002; Peri et al., 2007a) indicated that the LAI of the sward increased more slowly terms than the biomass when LAI was greater than 3 units (figure 12). This was consistent with Duru et al. (1997) who reported a unique exponential function for orchardgrass pasture between LAI and DM for different N levels. The non-linear relationship (mainly from LAI> 3) would be due to that the pseudo-stem length and weight increased with LAI (or time of regrowth) and consequently decreased the leaf:pseudo-stem ratio. Therefore, the increase in heavier pseudo-stem component and the relative decrease in the green leaf component over time give a greater proportion of DM with a smaller increase in LAI. The importance of this single relationship between DM yield and LAI is that it includes differences in the orchardgrass canopy due to changes over time in morphological aspects of the sward, such as tiller population, pasture height and leaf size caused by environmental and management factors. This relationship between DM production and LAI can then be used together with a canopy photosynthesis model to determine the foliage (LAI) increment for each day of growth.

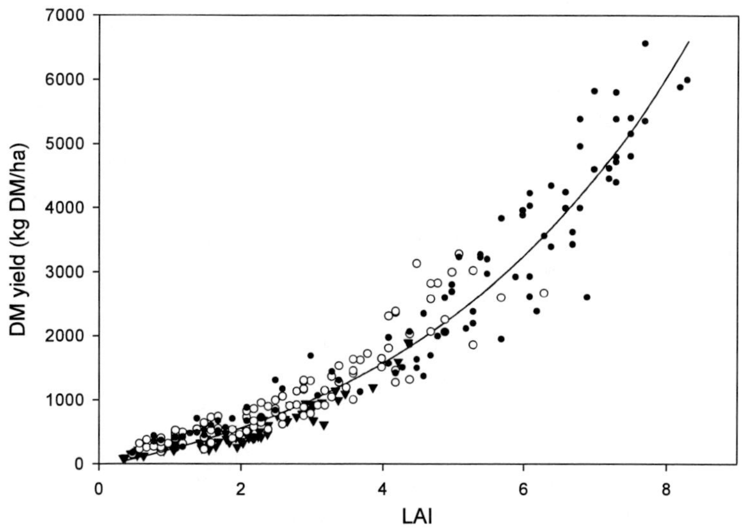

Figure 12. Accumulated dry matter (DM) yield (kg DM/ha) against leaf area index (LAI) for vegetative orchadgrass pastures. The line is for the fitted single exponential function. Observed data sorted by pasture under shade (▼), in open pastures (○) and with 300 kg N/ha (●) (Peri, 2002).

4. Modelling Orchardgrass Growth Based on Canopy Photosynthesis

A criticism of much of the agronomic work reported for orchardgrass and silvopastoral systems is that the influence of each of these factors (shade, temperature, N, water and regrowth duration) has usually been expressed in isolation or by their influence on seasonal production. There is limited explanation of the physiological basis for the responses, and consequently no predictive capacity for DM production. This limits the application of results to environments, sites and seasons outside those in which they were measured. To overcome this, an important research goal must be to predict pasture growth rates using a physiological basis and taking into account potential interactions between environmental and management factors.

A physiologically-based description of pasture growth operates through changes in the efficiency of conversion of energy to DM, and the total amount of energy available for this conversion. This is in turn influenced by the combination of light interception and the photosynthetic activity of individual leaves within the canopy, which are affected by environmental and management factors (Monteith, 1965; Sheehy and Cooper, 1973). Thus, when for example, nitrogen, water, temperature, or light are limited canopy photosynthesis may be restricted by both leaf area development and the leaf capacity for photosynthesis.

The predictive capability physiologically based pasture models makes them powerful tools for pasture management or in assisting agronomists to improve practices in pastoral or silvopastoral systems. They may also expose gaps in our knowledge at the sub-model level (i.e. the individual processes) such as photosynthesis (Thornley, 1998). Physiologically based canopy photosynthesis models, used to predict growth, are based on three main integrated components or sub-models: (i) the light intercepted by leaf surfaces (dependent upon LAI and canopy architecture) at different depths in the canopy, (ii) the resulting photosynthesis rate of those leaves, and (iii) subsequent partitioning of photosynthates to respiration (Wilson, 1960; Monteith, 1965; Duncan et al., 1967; Marshall and Biscoe, 1980a,b; Weir et al., 1984; Loomis and Connor, 1992; Thornley, 1998). The rate of canopy photosynthesis is derived by integrating the leaf photosynthetic rate throughout the depth of the canopy. The leaf photosynthetic rate, as a function of photosynthetic photon flux density (PPFD), is described by a non-rectangular hyperbola (Marshall and Biscoe, 1980a; Thornley, 1998) that has three parameters: the light-saturated rate which represents the asymptote or maximum saturated leaf photosynthetic rate (*Pmax*), the initial slope of the light response curve or photosynthetic efficiency (α) and a dimensionless parameter indicating the degree of curvature (θ).

The influence of environmental factors on orchardgrass net canopy photosynthesis (Pn) in pastoral (Peri et al., 2003c) and silvopastoral (Peri et al., 2006) systems for predicting DM production has been defined (figure 13). The model incorporate the leaf photosynthesis multiplicative sub-model developed for *Pmax* (Peri et al., 2002a,b; Peri et al., 2003a,b) and the 'law of the minimum factor' model for α (Peri et al., 2005a) when one or more environmental factor (temperature, water status, herbage N content and light regimes) was limiting. The range for predicting *Pmax* and α of orchardgrass leaves from the individual functions are: air temperatures from 2 to 37 °C, water status from ψ_{lp} –0.1 to –16.0 bar (corresponding to a soil VWC in the top 500 mm of 8.5 to 34%), foliage N content from 15 to 59 g N kg^{-1} DM, and duration of shade from 1 to 180 min. In non-limiting conditions *Pmax* had a value of 1.21 mg CO_2 m^{-2} s^{-1} and α of 0.0069 mg CO_2 J^{-1}. This model also has an interaction function [*f(TxN)*] for situations of low N ($\leq$ 20 g N kg^{-1} DM) and high temperature (> 23 °C) (Peri et al., 2002a), and an interaction function [*f(ShxW)*] between time under severe shade (5% of the open PPFD) and water stress (ψ_{lp} = -0.4 to –1.3 MPa) (Peri et al., 2003b). For θ, a constant mean value of 0.96 was used for all predictions (Peri et al., 2005a). In addition, Peri et al. (2003b) reported that air temperature was up to 7.4 °C higher than canopy temperature under tree shade for orchardgrass pasture grown in the same experimental site. For this reason, canopy temperature was estimated to predict *Pmax* and α for plants under severe shade using the exponential function proposed by Peri et al. (2003b). Non-limiting conditions for an individual factor were defined as: a constant temperature of 21 °C throughout the day; 5% N for herbage content; ψ_{lp}= -0.1 bar for leaf water status; and 20 days for regrowth duration.

In figure 14A, predicted *Pn* values for the three temperature regimes are presented (Peri et al., 2003c). In each case, the *Pn* response was parabolic against LAI and increased to reach a maximum and then declined as LAI increased further. The changes in air temperature affected the maximum value and shape of the *Pn* response. For the spring day (max. 21 °C) the maximum *Pn* was 30.8 g CO_2 m^{-2} d^{-1} at LAI= 3.8, and *Pn* was 16.3 g CO_2 m^{-2} d^{-1} at LAI= 9.5. In contrast, the maximum *Pn* for a summer day (max. 31 °C) was 15.5 g CO_2 m^{-2} d^{-1} at LAI= 2.3, and *Pn* was zero at LAI= 6.3. For the winter day (max. 10 °C), the maximum *Pn* value (18.3 g CO_2 m^{-2} d^{-1}) was reached at LAI= 5. The optimum range for Pn_{max} (19-22 °C) was smaller than that for Pmax (19-23 °C) of the youngest fully expanded leaf (Peri et al., 2002a). The difference between leaf and canopy photosynthesis was caused by maintenance respiration (Woledge and Dennis, 1982).

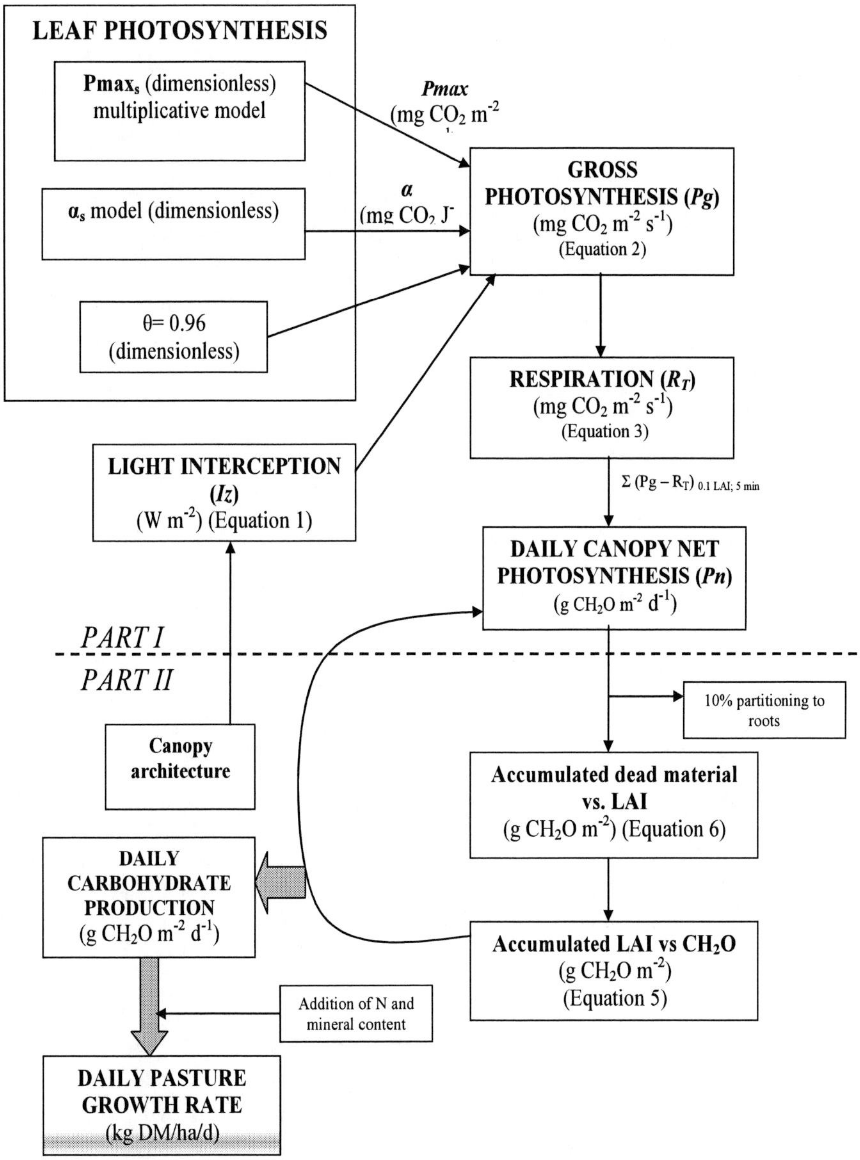

Figure 13. Generalised diagram of the canopy photosynthesis model (Peri et al., 2003c, 2006). Part *I* relates to simulations of net daily canopy photosynthesis (*Pn*) for orchardgrass in different environmental and management conditions calculated from canopy gross photosynthesis (*Pg*), light interception (*Iz*) and total respiration (R_T) every 0.1 units of LAI for each 5 minutes during a day. Part *II* (together with Part *I*) shows how daily pasture growth rate can be simulated from the canopy photosynthesis model for orchardgrass pastures grown under a diverse range of field conditions. $Pmax_s$ represents the standardised saturated leaf photosynthetic rate from the multiplicative model proposed by Peri et al. (2002a,b) and Peri et al. (2003a,b); α is the initial slope of the light response curve or the photosynthetic efficiency from the 'law of the minimum factor' model for α (Peri et al., 2005a); θ is a dimensionless parameter indicating the degree of curvature of the photosynthesis curve.

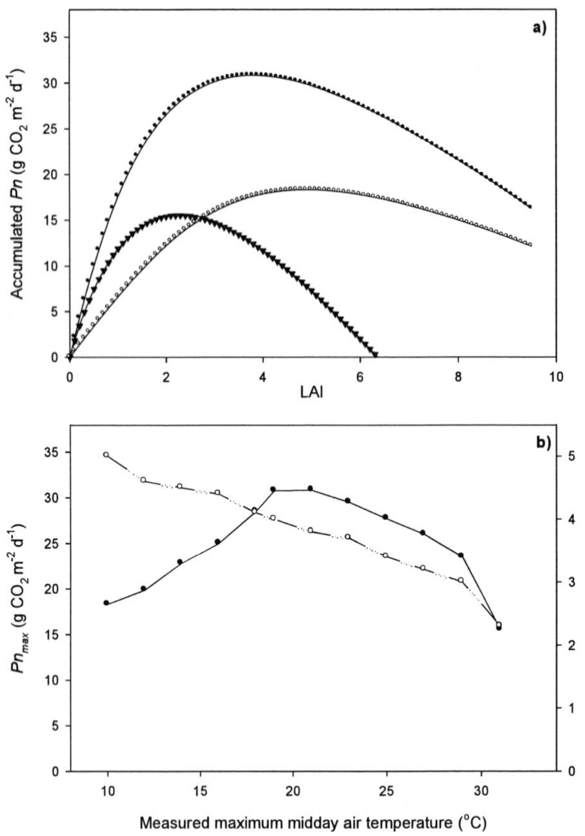

Figure 14. a) Predicted accumulated daily net canopy photosynthesis (*Pn*) against leaf area index (LAI) for measured changes in air temperature for a orchardgrass pasture where other factors were non-limiting. Simulations of *Pn* were analysed for maximum diurnal temperature of 21 °C (spring day) (●), 31 °C (summer day) (▼), and 10 °C (winter day) (○); b) predicted maximum daily net canopy photosynthesis (Pn_{max}) (●) and optimum leaf area index (LAI) (○) against measured maximum midday air temperature for a orchardgrass pasture where other factors were non-limiting (Peri et al., 2003c).

The decline in Pnmax with temperatures above 22 °C was probably caused by: (i) a decrease in Pmax and α due to an increase in the photorespiration rate with temperature (Peri et al., 2002a); and (ii) an increase in maintenance respiration with temperature. This was consistent with the results of Knievel and Smith (1973) who showed that temperatures above 28 °C greatly reduce orchardgrass growth.

The effect of the measured daily temperature regime on the maximum *Pn* (Pn_{max}) affected the optimum LAI (LAI at Pn_{max}) (figure 14b). Pn_{max} increased approximately linearly by 1.4 g CO_2 m^{-2} d^{-1} per °C, from 10 to 19 °C, then plateaued at 30.8 g CO_2 m^{-2} d^{-1} from 19 to 22 °C, and finally declined down to the lowest value. The optimum LAI declined with increasing air temperature from LAI= 5 at 10 °C to LAI= 2.3 at 31 °C. This is because the increase of maintenance respiration with temperature was proportional to LAI whereas the extra LAI deep in the canopy produced little photosynthesis. In practical terms, during summer regrowth periods, when maximum temperatures are often higher than 22 °C, the rotation length should be shorter (a lower LAI) to maximise canopy photosynthesis and avoid loss of DM by respiration.

Simulations also determined that *Pn* was affected by foliage N content due to a combination of; (i) the photosynthetic capacity of individual leaves was reduced (both *Pmax* and α) at low N levels and, (ii) canopy respiration increased with high levels of N. For orchardgrass, these effects combined showed an optimal value of 40 g N kg^{-1} DM (figure 15A). Thus, from 40 to 15 g N kg^{-1} DM content, Pn_{max} decreased by 67%. In contrast, at a leaf level, the measured *Pmax* response had a higher critical N content with a value of 52 g N kg^{-1} DM (Peri et al., 2002a). The difference arose because of the effect of foliage N on canopy maintenance respiration through an increase in the maintenance coefficient *b* determined that Pn_{max} decreased by 2.5% at 59 g N kg^{-1} DM compared with 40 g N kg^{-1} DM. Thus, the critical N content for net canopy photosynthesis was simulated to be lower than *Pmax*. This effect on maintenance respiration meant that foliage N content had a small effect on optimum LAI (figure 15B). Thus, respiration affected by N content had a small influence at LAI< 3 when Pn was maximum which indicates that a similar LAI should be aimed for grazing independently of the herbage N content.

Also Peri et al. (2003c) showed that Pn_{max} decreased in a negative curvilinear fashion with increasing water stress (Figure 16A). This was due to a greater decline in Pmax and α than in maintenance respiration. Thus, Pn_{max} decreased from 33.5 g CO_2 m^{-2} d^{-1} in irrigated plants to zero at a ψ_{lp}= -12.5 bar. The positive range for Pmax was greater in the youngest expanded leaf (ψ_{lp}= -0.1 to –14.0 bar) (Peri et al., 2002a). This was because a decrease in maintenance respiration with increasing water stress partially compensated for the reduction in Pmax. Jones et al. (1980) using a different measure of leaf water stress found a 50% field reduction in canopy photosynthesis for stressed plants. Similarly in controlled conditions, a sward which developed a rapid water stress reduced canopy photosynthesis from 10.0 g CO_2 m^{-2} h^{-1} in irrigated plants to only 2.2 g CO_2 m^{-2} h^{-1} in the severe water stress simulated sward. Therefore, while no direct comparison can be made between the data reported by Jones et al. (1980) and simulated values from the model proposed in this study the magnitudes of the effects are similar. In addition, the combined effect of low Pn and only a small reduction in R_T at severe water stress determined that the optimum LAI decreased sharply from ψ_{lp}= -11.5 to –13.0 bar (Figure 16B) which indicates that a high LAI can not be sustained under such conditions.

The net canopy photosynthesis rate of orchardgrass plants grown under high to low light intensities was predicted to decrease as a function of the intensity and time of the PPFD level experienced (Peri et al., 2005b, 2006). Specifically, Pn_{max} was predicted to decrease approximately linearly from 33.4 g CO_2 m^{-2} d^{-1} to zero as PPFD fell from full sunlight to 10% of open PPFD in a fluctuating light regime (figure 17A). Overall, Pn_{max} was predicted to be reduced depending on the duration of shade to which leaves were exposed. Thus, Pn_{max} of orchardgrass plants exposed to a fluctuating regime of 90% transmissivity was 97% higher than plants exposed to 20% transmissivity (figure 17A). This was mainly because the time at maximum *Pmax* (1.21 mg CO_2 m^{-2} s^{-1}) and α (0.0069 mg CO_2 J^{-1}) at 90% transmissivity was longer than for 20% transmissivity. Furthermore, the time required for full induction after shade for *Pmax* and α was predicted to be shortest for the individual leaves experiencing 90% transmissivity. A decrease in the photosynthetic capacity of leaves under shade would be responsible for the decrease in canopy photosynthesis. In this study, stomatal and non-stomatal factors were jointly responsible for the reduction and induction of *Pmax*, with their relative importance dependent on the duration and intensity of shade (Peri et al. 2002b). Also, the magnitude for reduction in α depended on the duration and intensity of shade (Peri et al. 2005a).

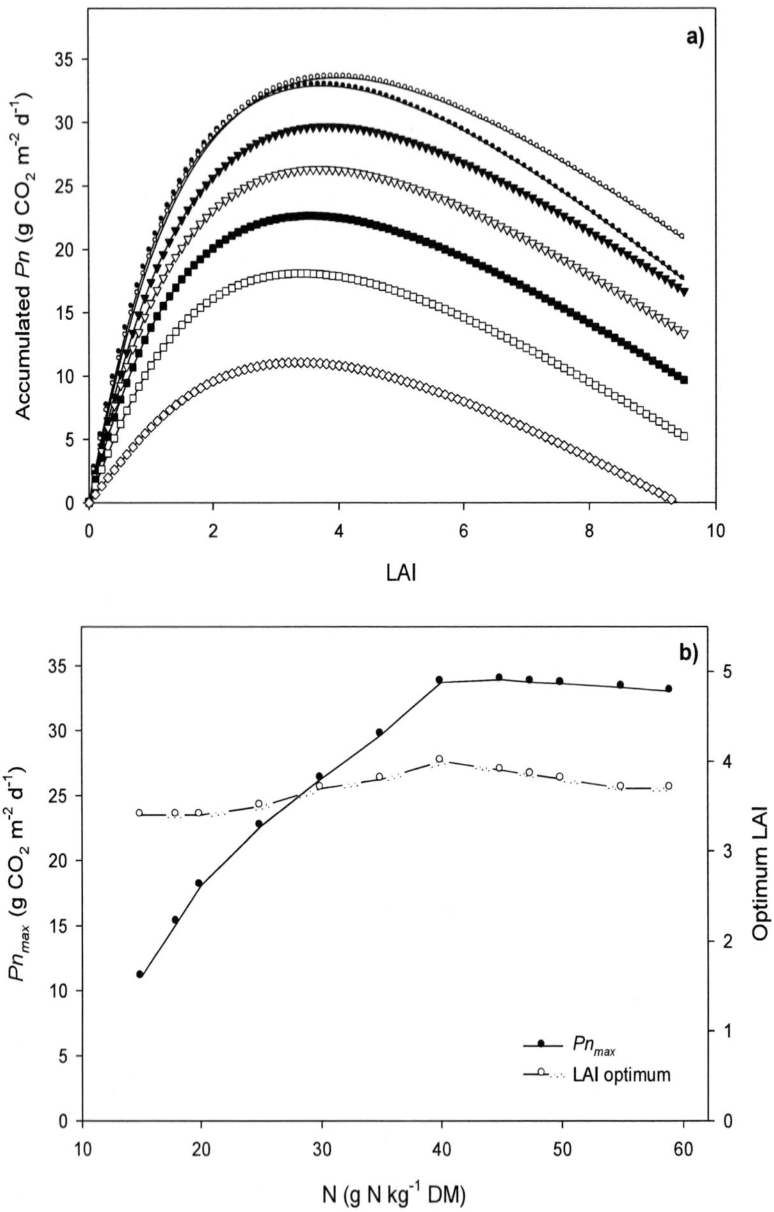

Figure 15. a) Predicted accumulated daily net canopy photosynthesis (*Pn*) against leaf area index (LAI) for different measured foliage nitrogen content: 59 (●), 40 (○), 35 (▼), 30 (▽), 25 (■), 20 (ϒ) and 15 (◊) g N kg^{-1} DM. b) predicted maximum daily net canopy photosynthesis (Pn_{max}) (●) and optimum leaf area index (LAI) (○) against measured foliage nitrogen content (g N kg^{-1} DM) for a orchardgrass pasture where other factors were non-limiting (Peri et al., 2003c).

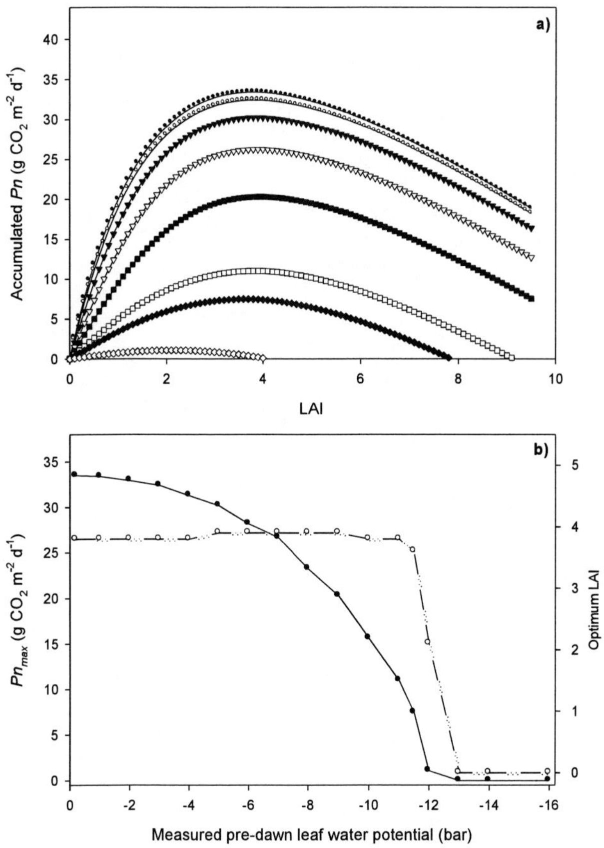

Figure 16. a) Predicted accumulated daily net canopy photosynthesis (*Pn*) against leaf area index (LAI) for different measured water status expressed as pre-dawn leaf water potential (ψ_{lp}): -0.2 (●), -3.0 (○), -5.0 (▼), -7.0 (▽), -9.0 (■), -11.0 (□), -11.5 (♦) and –12.0 (◇) bar. b) predicted maximum daily net canopy photosynthesis (Pn_{max}) (●) and optimum leaf area index (LAI) (○) against measured water status (ψ_{lp}) for a orchardgrass pasture where other factors were non-limiting (Peri et al., 2003c).

The simulated optimum LAI decreased linearly from 3.7 under a full sunlight regime to 0.7 under 20% transmissivity (figure 17B). This was because *Pn* decreased with a decline in daily PPFD due to the influence of maintenance respiration, which caused a decrease in the optimum LAI. The maintenance respiration also defined the level of *Pn* after the optimum LAI. For example, under full sunlight conditions *Pn* was 18.8 g CO_2 m^{-2} d^{-1} at LAI= 9.5, under a fluctuating light regime of 20% transmissivity *Pn* was zero at LAI= 1.6.

Also, it was predicted that the continuous light regime with 50% transmissivity had higher Pn_{max} than the same intensity under a fluctuating light regime (10.4 vs 8.4 g CO_2 m^{-2} d^{-1}) (figure 17A). This differential response would be caused by differences in the leaf photosynthetic rate (*Pmax* and *α*). The differential responses to intensity and duration of fluctuating light regimes means that, to accurately predict understorey responses of species in silvopastoral systems, or other fluctuating light regime conditions, slatted rather than cloth structures are required.

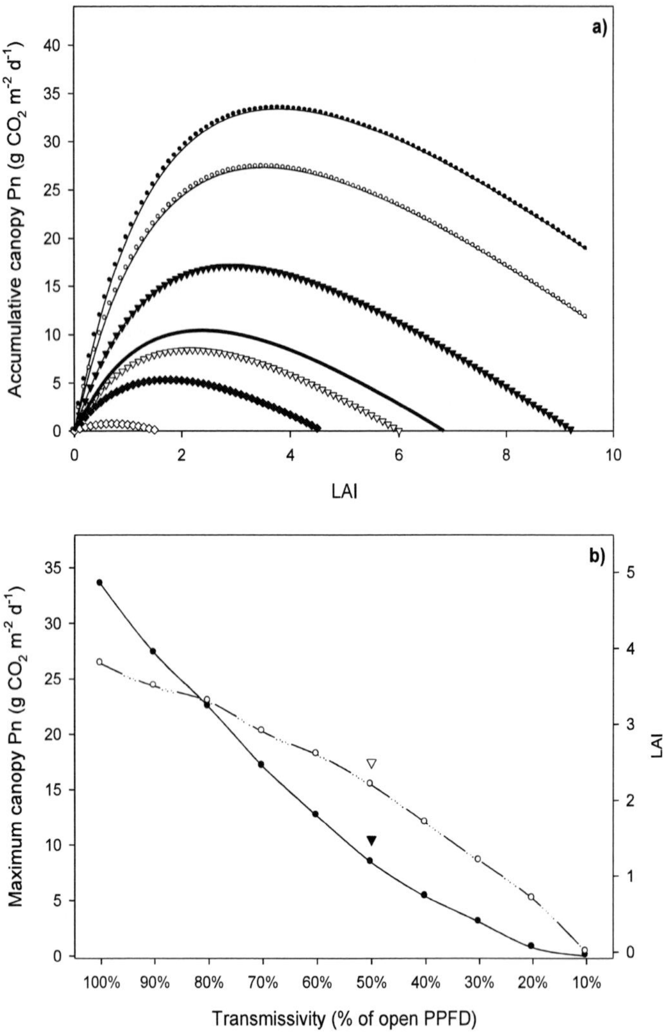

Figure 17. a) Predicted accumulated daily net canopy photosynthesis (*Pn*) against leaf area index (LAI) for different light regimes: full sunlight in the open (●), continuous cloth shade (▬) and fluctuating severe shade at different light intensities of 90 (○), 70 (▼), 50 (▽), 40 (♦) and 20% (◇) transmissivity) for a orchardgrass pasture where other factors were non-limiting; b) Predicted maximum daily net canopy photosynthesis (Pn_{max}) (●) and optimum leaf area index (LAI) (○) against different intensities of fluctuating light regime for a orchardgrass pasture when other factors were non-limiting. Pn_{max} (▼) and optimum LAI (▽) values for a continuous 50% transmissivity light regime are also indicated.

The canopy photosynthesis model was used successfully to predict orchardgrass DM production for a wide range of temperature, N and soil moisture conditions under fluctuating light regimes (Peri et al., 2003c; 2005a,b). The validation runs indicated approximately 85% of the variation in orchardgrass growth could be accounted for using the canopy photosynthesis model proposed. Thus, the model accurately simulated daily growth for the range of 6 to 134 kg DM ha^{-1} d^{-1} and for total production from 110 to 5100 kg DM ha^{-1}.

5. NUTRITIVE VALUE

Feeding value of pasture is the product of voluntary feed consumption and the digestibility of nutrients consumed. A limited number of studies can be found emphasizing effects of shading on nutritive value of temperate pastures. Most of these results have been generated by using continuous shading by cloth (Devkota et al., 1997). However, in silvopastoral systems understorey plants experience fluctuations in irradiance from full sun to shade. Peri et al. (2007a) reported that de crude protein (CP) percentage increased as PPFD declined with mean values ranged from 18.6% in open (100% transmissivity) to 22.5% under severe shade (~24% transmissivity) (figure 18). The increase in CP percentage as PPFD declined may be attributed to either a decrease in photosynthates, with a consequent rise in the nitrogen concentration, or to an increase in soil organic matter mineralisation under trees that provided more nitrogen for grass uptake (Wilson and Ludlow, 1991). The major effects of different light transmission intensity on the mean N concentration of six tropical grasses was in plants under moderate shade (68% light transmission) compared with pastures grown at 100% light transmission, but differences between species were much larger than any shade effects observed in those studies (Norton et al., 1991). The CP percentage with shade under trees showed less difference during severe drought (soil moisture < 18%) in summer (figure 18) with less difference between open and shaded treatments. Similarly Whitehead (1995) reported that dry conditions may retard both pasture uptake and soil mineralisation of N during the season.

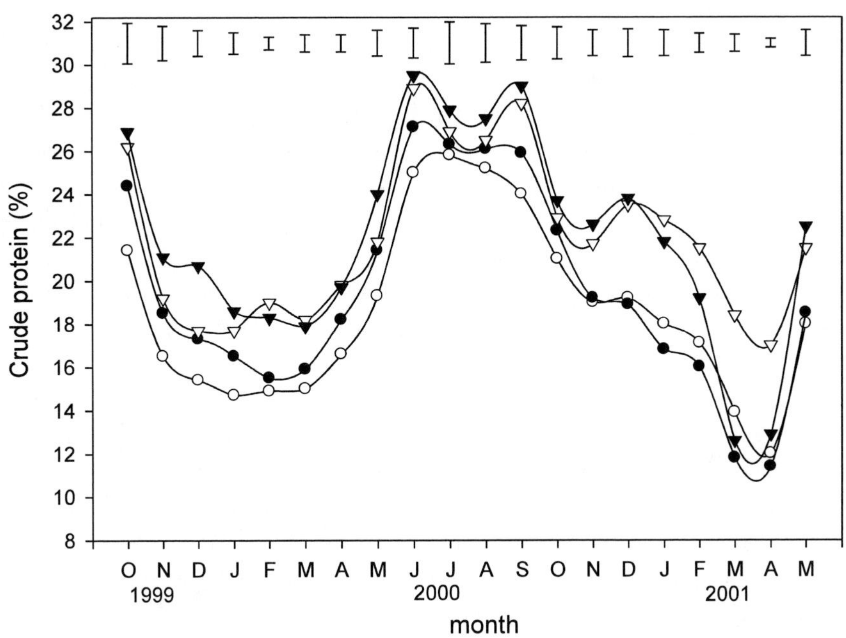

Figure 18. Orchardgrass crude protein (CP%) content over time for four shade treatments at the Lincoln University silvopastoral experiment (Canterbury, New Zealand): open (○) (100% transmissivity), open+slats (▽) (~43% transmissivity), under trees (●) (~58% transmissivity) and trees+slats (▼) (~24% transmissivity). Bars indicate standard error of the mean (sem) (Peri et al., 2007a).

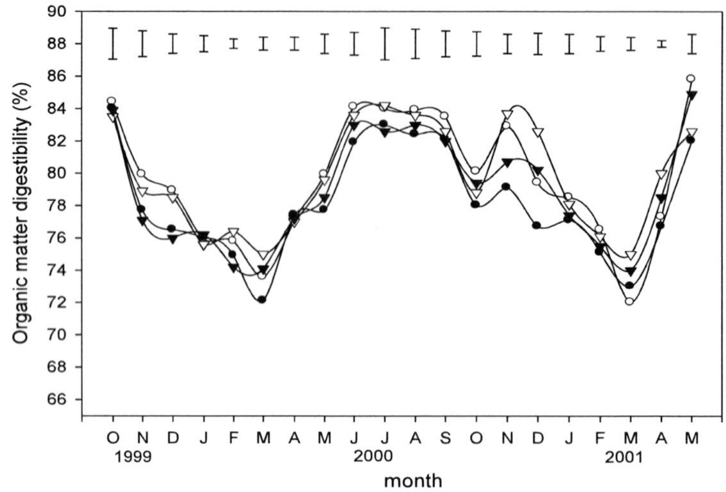

Figure 19. Orchardgrass organic matter digestibility (OMD%) over time for four shade treatments at the Lincoln University silvopastoral experiment (Canterbury, New Zealand): open (○) (100% transmissivity), open+slats (▽) (~43% transmissivity), under trees (●) (~58% transmissivity) and trees+slats (▼) (~24% transmissivity). Bars indicate standard error of the mean (sem) (Peri et al., 2007a).

In contrast, the intensity of fluctuating shade had little effect on organic matter digestibility (OMD) with a mean value of 79 ±3.2% (Peri et al., 2007a), but with a fluctuating seasonal variation which ranged from 72.0 to 85.8 % (figure 19). Maximum values corresponded to spring, winter and autumn regrowth and minimum values when orchardgrass pastures were under water stress (soil VWC < 18%). There were small differences between open and shaded treatments for organic matter digestibility (OMD) (figure 19). Many results have shown that in vitro digestibility of grasses is reduced under low light intensity and this is associated with reduced digestibility of cell wall constituents (Hight et al., 1968). However, other studies have obtained less conclusive results with both increases and decreases in digestibility of herbage grown in a shaded environment (Wilson, 1988). Cell walls are the major fraction available for digestion and it is important to determine whether shade affects their composition (Norton et al., 1991). Shading usually reduces the total non-structural carbohydrate of grasses, but has variable (positive and negative) effects on cell wall content and composition, lignin and in vitro digestibility of plant dry matter (Wilson, 1988). According to Hight et al. (1968), soluble carbohydrate content decreased by 3.7% units and dried forage digestibility by 0.6-3.6% units during a short shading period of time (2-3 days at 22 % light transmission). However, Norton et al. (1991) found a small non-significant decrease in soluble carbohydrate and a decrease in *in vitro* dry matter digestibility.

Table 4 indicates that mean macro-nutrient concentrations of orchardgrass leaves increased as the PPFD level declined, mainly between the open (100% transmissivity) and the ~24% transmissivity condition (P 4.2 vs 5.2, Mg 1.6 vs 2.1, K 29.4 vs 33.9, Ca 4.7 vs 5.3, S 2.6 vs 3.6 g kg^{-1} DM). One reason for this may be because heavily shaded pasture (i.e. less than 30-40% light transmission) has low dry matter contents or higher water contents (Reynolds, 1995). Burton et al. (1959) reported that severe shade reduced the yield of DM by about one-third, the concentration of P, Ca and Mg in Coastal Bermudagrass increases by 25-50%. Furthermore, active mineralisation of soil organic matter by pine roots or their mycorrhiza was suggested as a likely mechanism accounting for elevated inorganic and available nutrient levels under pines in some New Zealand grassland soils and therefore more nutrient availability for pastures in a silvopastoral system (Davis, 1994). Hawke and O'Connor (1993) reported that while soil potassium remained relatively constant (average of 7.8 ppm at 400 sph), magnesium, sulphate-sulphur and phosphorus levels tended to increase at higher tree stocking rates (14 mg kg^{-1}, 20 mg kg^{-1} and 50 mg kg^{-1}, respectively under an 18 year-old radiata pine stand). These changes in topsoils were attributed to nutrient uptake, organic matter mineralisation, and the effect of organic acids under the radiata pine stand.

However, there was an interaction between light intensity and time (seasons) with shade under trees which showed less difference during severe drought. Thus, macro-nutrient concentrations were higher (except Ca) in spring and winter (soil moisture > 24%) compared with summer (soil moisture < 18%). In general, water stress may reduce the capacity of roots to absorb nutrients if it reduces transpiration and growth, and therefore plant demand for nutrients (Lambers et al., 1998). While concentration of P in grasses is reduced by drought, the effect on other elements is less consistent (Rahman et al., 1971). The increase in Ca concentration in summer for the drought-resistant cultivar KM2 orchardgrass population could reflect a greater density of roots in deeper soil horizons (Volaire and Thomas, 1995).

Table 4. Mean mineral content (g kg^{-1} DM) in herbage orchardgrass pasture for different seasons at 21 days regrowth (2 years data) at the Lincoln University silvopastoral experiment (Canterbury, New Zealand). Open: 100% transmissivity, open+slats: ~43% transmissivity, trees: ~58% transmissivity and trees+slats: ~24% transmissivity (Peri et al., 2007a)

Date	Treatment	P	Mg	K	Ca	S
Spring	Open	4.6	1.5	32.6	4.0	2.5
	Open + Slats	4.7	1.7	34.4	3.9	3.6
	Trees	5.3	1.7	37.7	4.2	3.7
	Trees + Slats	5.6	1.9	40.1	4.3	3.3
Summer	Open	3.8	1.7	30.5	5.2	2.4
	Open + Slats	4.1	1.8	30.0	4.8	3.2
	Trees	5.2	2.0	32.2	5.5	3.6
	Trees + Slats	5.6	2.1	34.6	5.6	3.5
Autumn	Open	4.1	1.6	26.2	4.9	2.4
	Open + Slats	4.2	1.8	29.2	4.8	3.0
	Trees	4.1	2.1	25.0	5.6	3.3
	Trees + Slats	4.2	2.1	29.0	5.2	3.2
Winter	Open	4.4	1.7	26.7	5.0	3.5
	Open + Slats	4.8	1.8	28.6	5.4	3.4
	Trees	5.5	2.0	29.0	5.8	3.7
	Trees + Slats	5.5	2.3	30.1	6.5	3.5
Mean annual	Open	4.2	1.6	29.4	4.7	2.6
	Open + Slats	4.4	1.8	30.8	4.6	3.3
	Trees	4.9	2.0	31.3	5.2	3.6
	Trees + Slats	5.2	2.1	33.9	5.3	3.3
	lsd	***0.60***	***0.11***	***2.45***	***0.36***	***0.27***

Macro-nutrient concentrations increased as PPFD declined, but the minimum values measured mainly in the open treatment (100% transmissivity) satisfy the demand for optimum animal feed requirements. According to Kemp et al. (1999) the optimum concentration of mineral elements in temperate pasture species for sheep growth are P= 2.0, S= 1.15, K= 3.6, Ca= 2.9 and Mg= 1.2 g kg^{-1} DM. However, according to Kemp et al. (1999) leaf nutrient concentrations (table 4) were below the optimum for pasture growth mainly for the open treatment and during summer drought (P= 4.0, S= 3.2, K= 25.0, Ca= 3.0 and Mg= 2.0 g kg^{-1} DM).

Given CP percentage and macro-nutrient concentrations (higher than the optimum required for sheep growth) increased slightly with shade level, digestibility values were not greatly influenced by shade, it seems likely that shaded orchardgrass pastures would limit animal production primarily due to lower DM production rates. This may lower pre-grazing pasture mass and reduce bulk density and bite size from the etiolated pasture. This is consistent with Peri *et al* (2001b) who reported that apparent intakes were reduced by 0.3 kg DM hd^{-1} d^{-1} for lambs grazing orchardgrass pastures under trees. This was attributed to a reduction in pasture bulk densities. Thus, grazing animals had greater potential intake in open pastures than under trees because the amount harvestable in a single bite for livestock grazing lower density pasture is smaller (Gong et al., 1996). Reduction in bite size leads to reduced daily pasture intake and live weight gain (LWG) because the animal normally cannot sufficiently compensate by increasing the rate of biting or grazing time per day. Alternatively, long grazing rotations could be used under shaded pastures to provide time to accumulate sufficient pasture mass. However, problems from longer spelling times would be: i) taller pasture and increased reproductive development that results in reduced bulk density, and ii) older forage of lower nutritive value.

REFERENCES

Alberda T. 1962. Actual and potential production of agricultural crops. *Netherlands Journal of Agricultural Science*, 10: 325-333.

Anderson GW. 1978. Productivity of crops and pastures under trees. In: KMW Howes and RA Rummery (Eds) *Integrated Agriculture and Forestry*, pp. 59-63. Perth: CSIRO, Division of Land Resource Management.

Auda H, Blaser RE, Brown RH. 1966. Tillering and carbohydrate contents of orchardgrass as influenced by environmental factors. *Crop Science*, 6: 139-143.

Belanger G. 1998. Morphogenetic characteristics of timothy grown with varying N nutrition. *Canadian Journal of Plant Science*, 78: 103-108.

Braziotis DC, Papanastasis VP. 1995. Seasonal changes of understorey herbage yield in relation to light intensity and soil moisture content in a Pinus pinaster plantation. *Agroforestry Systems*, 29: 91-105.

Brougham RW. 1956. Effect of intensity of defoliation on regrowth of pasture. *Australian Journal of Agricultural Research*, 7: 377-387.

Burton GW, Jackson JE, Knox FE. 1959. The influence of light reduction upon the production, persistence and chemical composition of Coastal Bermudagrass, *Cynodon dactylon*. *Agronomy Journal*, 51: 537-542.

Casal JJ, Deregibus VA, Sanchez RA. 1987. Tillering responses of *Lolium multiflorum* plants to changes of red/far-red ratio typical of sparse canopies. *Journal of Experimental Botany*, 38: 1432-1439.

Davies A. 1988. The regrowth of grass swards. The grass plant – its form and function. In: MB Jones and A Lazenby (Eds) *The Grass Crop: The physiological basis of production*. pp. 85-127. London: Chapman and Hall.

Davis MR. 1994. Topsoil properties under tussock grassland and adjoining pine forest in Otago, New Zealand. *New Zealand J Agr Research*, 37: 465-469.

Devkota NR, Kemp PD, Hodgson J. 1997. Screening pasture species for shade tolerance. *Proceedings of the Agronomy Society of New Zealand*, 27: 119-128.

Devkota NR, Kemp PD, Valentine I, Hodgson J. 2000. Shade tolerance of pasture species in relation to deciduous tree, temperate silvopastoral systems. *Proceedings of the Agronomy Society of New Zealand*, 30, 101-107.

Duncan WG, Loomis RS, Williams WA, Hanau R. 1967. A model for simulating photosynthesis in plant communities. *Hilgardia*, 38: 181-205.

Duru M, Lemaire G, Cruz P. 1997. Grasslands. In: G Lemaire (Ed) *Diagnosis of the Nitrogen Status in Crops*, pp. 59-72. New York: Springer.

Duru M, Ducrocq H. 2000. Growth and senescence of the successive leaves on a cocksfoot tiller. Effect of nitrogen and cutting regime. *Annals of Botany*, 85: 645-653.

Eagles C.F. 1967. The effect of temperature on vegetative growth in climatic races of *Dactylis glomerata* in controlled environments. *Annals of Botany*, 31: 31-39.

Eagles CF, Treharne KJ. 1969. Photosynthetic activity of *Dactylis glomerata* L. in different light regimes. *Photosynthetica*, 3: 29-38.

Evans JR. 1996. Developmental constraints on photosynthesis: Effects of light and nutrition. In: NR Baker (Ed) *Photosynthesis and the Environment*, pp. 281-304. Dordrecht: Kluwer Academic Publisher.

Field C. 1983. Allocating leaf nitrogen for the maximization of carbon gain: Leaf age as a control on the allocation program. *Oecologia*, 56: 341-347.

Frank AB, Barker RE. 1976. Rates of photosynthesis and transpiration and diffusive resistence of six grasses grown under controlled conditions. *Agronomy Journal*, 68: 487-490.

Garnier E, Roy J. 1988. Modular and demographic analysis of plant leaf area in sward and woodland populations of *Dactylis glomerata* and *Bromus erectus*. *Journal of Ecology*, 76: 729-743.

Gong Y, Hodgson J, Lambert MG, Gordon IL. 1996. Short-term ingestive behaviour of sheep and goats grazing grasses and legumes. 2. Quantitative relationships between sets of sward and ingestive behaviour variables. *New Zealand J Agr Research*, 39: 75-82.

Hawke MF, O'Connor MB. 1993. Soil pH and nutrient levels at Tikitere agroforestry research area. *New Zealand J For Science*, 23: 40-48.

Hight GK, Sinclair DP, Lancaster RJ. 1968. Some effects of shading and of nitrogen fertiliser on the chemical composition of freeze-dried and oven-dried herbage, and on the nutritive value of oven-dried herbage fed to sheep. *New Zealand J Agr Research*, 11: 286-302.

Hilbert DW, Larigauderie A, Reynolds J. 1991. The influence of carbon dioxide and daily photon-flux density on optimal leaf nitrogen concentration and root:shoot ratio. *Annals of Botany*, 68: 365-376.

Hirose T, Werger MJA. 1987. Maximizing daily canopy photosynthesis with respect to the leaf nitrogen allocation pattern in the canopy. *Oecologia*, 72: 520-526.

Hsiao TC, Acevedo E. 1974. Plant responses to water deficit, water use efficiency, and drought resistance. *Agricultural Meteorology*, 14: 59-84.

Jackson DK. 1974. The course and magnitude of water stress in *Lolium perenne* and *Dactylis glomerata*. *Journal of Agricultural Science, Cambridge*, 82: 19-27.

Jones MB, Leafe EL, Stiles W. 1980. Water stress in field-grown perennial ryegrass. I. Its effect on growth, canopy photosynthesis and transpiration. *Annals of Applied Biology*, 96: 87-101.

Joshi MR, Lucas RJ, Sedcole JR, Pollock KM, Moot DJ, Mead DJ. 1999. Shading effects of Pinus radiata on productivity and feeding value of orchardgrass pasture. *Agroforestry Forum*, 9: 17-19.

Kemp PD, Condron LM, Matthew C. 1999. Pastures and soil fertility. In: J White, J Hodgson (Eds) New Zealand Pastures and Crop Science. Oxford University Press, Auckland.

Knapp AK, Smith WK. 1987. Stomatal and photosynthetic responses during sun/shade transitions in subalpine plants: influence on water use efficiency. *Oecologia*, 74: 62-67.

Knievel DP, Smith D. 1973. Influence of cool and warm temperatures and temperature reversal at inflorescence emergence on growth of timothy, orchardgrass, and tall fescue. *Agronomy Journal*, 65: 378-383.

Lambers H, Chapin III FS, Pons TJ. 1998. Plant Physiological Ecology. Springer-Verlag, New York.

Lancashire JA, Brock JL. 1983. Management of new cultivars for dryland. *Proceeding of the New Zealand Grassland Association*, 44: 61-73.

Lawlor DW. 1972. Growth and water use of *Lolium perenne* II. Plant growth. *Journal of Applied Ecology*, 9: 99-105.

Lemaire G. 1997. Diagnosis of the Nitrogen Status in Crops. Springer Verlag, Berlin.

Lemaire G, Salette J, Gosse G, Chartier M. 1983. Temperature and spring growth for grasses: Comparison of different grass stands. In: AJ Corral (Ed) Efficient Grassland Farming, Occasional Symposium of the British Grassland Society, 14: 255-261.

Loomis RS, Connor DJ. 1992. *Crop Ecology: Productivity and Management in Agricultural Systems*. Melksham: Cambridge University Press.

Marshall B, Biscoe PV. 1980a. A model for C_3 leaves describing the dependence of net photosynthesis on irradiance. I. Derivation. *Journal of Experimental Botany*, 31: 29-39.

Marshall B, Biscoe PV. 1980b. A model for C_3 leaves describing the dependence of net photosynthesis on irradiance. II. Application to the analysis of flag leaf photosynthesis. *Journal of Experimental Botany*, 31: 41-48.

Miah MG, Garrity DP, Aragon ML. 1995. Light availability to the understorey annual crops in an agroforestry system. In: H Sinoquet and P Cruz (Eds) Ecophysiology of Tropical Intercropping, pp. 99-107. Paris: Institute National de la Recherche Agronomique.

Mills A, Moot DJ, Jamieson PD. 2009. Quantifying the effect of nitrogen on productivity of cocksfoot (*Dactylis glomerata* L.) pastures. *European Journal of Agronomy*, 30: 63-69.

Mitchell KJ, Lucanus R. 1960. Growth of pastures species under controlled environments. *New Zealand Journal of Agriculture Research*, 3: 647-665.

Moloney SC. 1993. Selection, management and use of cocksfoot cultivars in North Island pastoral farming. *Proceedings of the New Zealand Grassland Association*, 55: 119-125.

Monteith JL. 1965. Light distribution and photosynthesis in field crops. *Annals of Botany*, 29: 17-37.

Monteith J.L. 1977. Climate and the efficiency of crop production in Britain. *Philosophical Transactions of the Royal Society of London, -B*, 281: 277-294.

Nie GY, Long SP, Baker NR. 1992. The effects of development at sub-optimal growth temperatures on photosynthetic capacity and susceptibility to chilling-dependent photo-inhibition in *Zea mays*. *Physiologia Plantarum,* 85: 554-560.

Norris IB. 1982. Soil moisture and growth of contrasting varieties of *Lolium*, *Dactylis* and *Festuca* species. *Grass and Forage Science*, 3: 273-283.

Norton BW, Wilson JR, Shelton HM, Hill KD. 1991. The effect of shade on forage quality. In: HM Shelton, WW Stur (Eds) Forages for Plantation Crops. ACIAR, Canberra.

Pearce RB, Brown RH, Blaser RE. 1965. Relationships between leaf area index, light interception and net photosynthesis in orchardgrass. *Crop Science*, 5: 553-556.

Penman HL. 1962. Woburn Irrigation, 1951-59. II. Results for grass. *Journal of Agricultural Science, Cambridge*, 58: 349-364.

Peri PL, Lucas RJ, Moot DJ, Varella AC, McNeil DL. 2001a. Optimising yield and quality of orchardgrass pasture in temperate silvopastoral systems. *Proceedings XIX International Grassland Congress*, pp. 657-658. San Pablo, Brasil, 2001.

Peri PL, Varella AC, Lucas RJ, Moot DJ. 2001b. Cocksfoot and lucerne productivity in a *Pinus radiata* silvopastoral system: a grazed comparison. *Proceedings of the New Zealand Grassland Association*, 63: 139-147.

Peri PL, Varella AC, Lucas RJ, Moot DJ. 2001b Cocksfoot and lucerne productivity in a *Pinus radiata* silvopastoral system: a grazed comparison. *Proceedings of the New Zealand Grassland Association*, 63: 139-147..

Peri PL. 2002. Leaf and canopy photosynthesis models for cocksfoot (*Dactylis glomerata* L.) grown in a silvopastoral system. Ph.D thesis, Lincoln University, Canterbury, New Zealand. 291 pp.

Peri PL, Moot DJ, McNeil DL, Varella AC, Lucas RJ. 2002a. Modelling net photosynthesis rate of field grown cocksfoot leaves under different nitrogen, water and temperature regimes. *Grass and Forage Science*, 57: 61-71.

Peri PL, McNeil DL, Moot DJ, Varella AC, Lucas RJ. 2002b. Net photosynthesis rate of cocksfoot leaves under continuous and fluctuating shade conditions in the field. *Grass and Forage Science*, 57: 157-170.

Peri PL, Lucas RJ, Moot DJ. 2002c. Urine patches indicate yield potential of cocksfoot. *Proceedings of the New Zealand Grassland Association*, 64: 73-80.

Peri PL, Moot DJ, McNeil DL, Lucas RJ. 2003a. Modelling net photosynthetic rate of field-grown cocksfoot leaves to account for regrowth duration. *New Zealand Journal of Agricultural Research*, 46: 105-115.

Peri PL, Moot DJ, McNeil DL. 2003b. An integrated model for predicting maximum net photosynthetic rate of cocksfoot (*Dactylis glomerata*) leaves in silvopastoral systems. *Agroforestry Systems*, 58: 173-183.

Peri PL, Moot DJ, McNeil DL. 2003c. A canopy photosynthesis model to predict the dry matter production of cocksfoot pastures under varying temperature, nitrogen and water regimes. *Grass and Forage Science*, 58: 416-430.

Peri PL, Moot DJ, McNeil DL. 2005a. Modelling photosynthetic efficiency (α) for the light-response curve of cocksfoot leaves grown under temperate field conditions. *European Journal of Agronomy*, 22: 277-292.

Peri PL, Moot DJ, Lucas RJ, McNeil D. 2005b. *Prediction of canopy photosynthesis for cocksfoot pastures grown under different light regimes. Proceedings XX International Grassland Congress, Grass and forage physiology*, pp. 199. Dublin, Ireland, 2005.

Peri PL, Moot DJ, McNeil DL. 2006. Validation of a canopy photosynthesis model for cocksfoot pastures grown under different light regimes. *Agroforestry Systems*, 67: 259-272.

Peri PL, Lucas RJ, Moot DJ. 2007a. Dry matter production, morphology and nutritive value of *Dactylis glomerata* growing under different light regimes. *Agroforestry Systems*, 70: 63-79.

Peri PL, Moot DJ, Jarvis P, McNeil DL, Lucas RJ. 2007b. Morphological, anatomical, and physiological changes of orchardgrass leaves grown under fluctuating light regimes. *Agronomy Journal*, 99: 1502-1513.

Rahman AAA, Shalaby AF, El Monayeri MO. 1971. Effect of moisture stress on metabolic products and ion accumulation. *Plant Soil*, 34 :65-90.

Reynolds SG. 1995. Pasture-Cattle-Coconut Systems. FAO, RAPA (Regional Office for Asia and the Pacific), Bangkok.

Rinne K. 1978. The effect of maturity on herbage pasture during grazing season. *Annales Agriculturae Fenniae*, 17: 10-17.

Robson MJ, Parsons AJ. 1978. Nitrogen deficiency in small closed communities of S24 ryegrass. I. Photosynthesis, respiration, dry matter production and partition. *Annals of Botany*, 42: 1185-1197.

Robson MJ, Ryle GJA, Woledge J. 1988. The grass plant – its form and function. In: MB Jones and A Lazenby (Eds) The Grass Crop, pp. 25-83. London: Chapman and Hall.

Sheehy JE, Cooper JP. 1973. Light interception, photosynthetic activity, and crop growth rate in canopies of six temperate forage grasses. *Journal of Applied Ecology*, 10: 239- 250.

Sheehy JE. 1977. Microclimate, canopy structure and photosynthesis in canopies of three contrasting temperate forage grasses. III. Canopy photosynthesis, individual leaf photosynthesis and the distribution of current assimilate. *Annals of Botany*, 41: 593-604.

Sinclair TR, Muchow RC. 1999. Radiation use efficiency. *Advances in Agronomy*, 65: 215-265.

Smith H. 1982. Light quality, photoperception, and plant strategy. *Annual Review of Plant Physiology*, 33: 481-518.

Smith LC, Morton JD, Catto WD, Trainor KD. 2000. Nitrogen responses on pastures in the southern South Island of New Zealand. *Proceedings of the New Zealand Grassland Association*, 62: 19-23.

Stevens DR, Baxter GS, Stewart MJ, Casey MJ, Miller KB. 1992. Grasslands Kara cocksfoot: a productive cultivar under lax grazing. *Proceedings of the New Zealand Grassland Association*, 54: 143-146.

Thornley JHM. 1998. *Grassland Dynamics: An Ecosystem Simulation Model*. Wallingford: CAB International.

Treharne KJ, Cooper JP, Taylor TH. 1968. Growth response of orchardgrass (*Dactylis glomerata* L.) to different light and temperature environments. II. Leaf age and photosynthetic activity. *Crop Science*, 8: 441-445.

Treharne KJ, Eagles CF. 1970. Effect of temperature on photosynthetic activity of climatic races of *Dactylis glomerata* L. *Photosynthetica*, 4: 107-117.

van Keulen H, Goudriaan J, Seligman NG. 1989. Modelling the effects of nitrogen on canopy development and crop growth. In: G Russel, B Marshall and PG Jarvis (Eds) Plant canopies, Their Growth, Form and Function, pp. 83-104. Cambridge: Cambridge University Press.

Volaire F, Thomas H. 1995. Effects of drought on water relations, mineral uptake, water-soluble carbohydrate accumulation and survival of two contrasting population of cocksfot (*Dactylis glomerata* L.). *Annals of Botany*, 75: 513-524.

Weir AH, Bragg PL, Porter JR, Rayner JH. 1984. A winter wheat crop simulation model without water or nutrient limitations. *Journal of Agricultural Science, Cambridge*, 102: 371-382.

Whitehead DC. 1995. Grassland Nitrogen. Wallingford: CAB International.

Wilman D, Wright PT. 1983. Some effects of applied nitrogen on the growth and chemical composition of temperate grasses. *Herbage Abstracts*, 53: 387-393.

Wilson JW. 1960. Inclined point quadrats. *New Phytologist*, 59: 1-8.

Wilson JR. 1988. Ecological constraints to production and nutritive quality of pastures under tree crops. In: Livestock-Tree Cropping Production Systems. Proceedings FAO/MARDI International Workshop, Serdang, Malaysia.

Wilson JR, Ludlow MM. 1991. The environment and potential growth of herbage under plantations. In: HM Shelton and WW Stür (Eds) *Forages for Plantation Crops*, 32: 10-24.

Woledge J. 1972. The effect of shading on the photosynthesis rate and longevity of grass leaves. *Annals of Botany*, 36: 551-561.

Woledge J, Leafe EL. 1976. Single leaf and canopy photosynthesis in a ryegrass sward. *Annals of Botany*, 40: 773-783.

Woledge J, Dennis WD. 1982. The effect of temperature on photosynthesis of ryegrass and white clover leaves. *Annals of Botany*, 50: 25-35.

In: Grasslands: Types, Biodiversity and Impacts
Editor: Wen-Jun Zhang ISBN: 978-1-61470-555-0

Chapter 5

GRASSLAND MANAGEMENT AND STRUCTURAL CHANGES IN SOIL MICROBIAL COMMUNITIES

Zabed Hossain * ***and Shu-ichi Sugiyama***
Faculty of Agriculture and Life Science, Hirosaki University, Japan

ABSTRACT

Grasslands are complex terrestrial ecosystems where interactions among producers (green plants), consumers (grazing animals) and decomposers (soil microorganisms) determine its productivity and functions. Although a number of management activities are practiced in the grassland systems to maximize the efficiency of conversion of solar energy or plant mass into animal product, effects of managements on structure and functions of the below-ground communities has not been well studied. The present chapter discusses about the effects of management history on plant and soil microbial communities in grasslands by comparing two types of semi-natural grasslands (tall-type by infrequent grazing and short-type by frequent grazing) and the improved grassland in northern part of Japan. Plant species richness was highest in the tall-type grassland and lowest in the improved grassland. Short-type and tall-type grasslands were dominated by the proportions of C_4 and forbs species, respectively, while the improved grassland was by C_3 species. Microbial communities were studied by phospholipid fatty acid (PLFA) profiling and rDNA finger printing methods. Major soil microbial communities including total microbial PLFA, mycorrhizal PLFA, and the saprophytic fungal PLFA showed significant differences among the three grassland types. Fungal DNA band number also differed significantly among the grasslands. However, significant differences in community composition of microbial groups mostly appeared between semi-natural tall-type and the improved grasslands indicating the role of human management on structuring soil microbial communities. All these results also suggest that a complex interaction of human management and grazing might have altered the structure of plant and soil microbial communities in grasslands.

Keywords Grasslands, management, phospholipid fatty-acids (PLFA), rDNA, soil microbial community, vegetation

---

* Correspondence address: [1]Current address: Department of Plant Biology and Forest Genetics, Swedish University of Agricultural Science (SLU), Uppsala BioCenter, Uppsala 7080, Sweden. E-mail address: zabed@univ dhakaedu, sugi@hirosaki-u.ac.jp

1. INTRODUCTION

Plant and soil microbial communities in the terrestrial ecosystems are dependent on each other (Wardle 2002). Plants as producer provide the decomposer community with organic substances and nutrients through litter and root exudates. The decomposer community, in turn, influences plant communities by recycling dead plant material and making mineral nutrients available for plant use. Although the functional linkages between plant and soil microbes are well recognized, structural linkages between these two communities are not well clarified (Sugiyama et al., 2008; De Deyn and van der Putten, 2005; Hooper et al., 2005).

Grasslands are complex terrestrial ecosystems where interactions among producers (green plants), consumers (grazing animals) and decomposers (soil microorganisms) determine its productivity and functions. Grasslands constitute one thirds of the area of the terrestrial ecosystems (Schlesinger 1997) and support uniquely large and active soil microbial biomass (Bardgett and Cook, 1998). Therefore, any changes in plant and soil microbial communities are likely to influence ecosystem properties including productivity and nutrient cycling. Although a number of management activities are practiced in the grassland systems to maximize the efficiency of conversion of solar energy or plant mass into animal product, the consequences of such changes for the components of grasslands are not well studied. Understanding the effects of management on plant and soil microbial communities is, thus, relevant not only for the proper management but also for enhancing our knowledge about the structure and function of the grassland ecosystems.

Over the decades, study of soil microbial communities has been hampered due to the tremendous diversity of microbes in soil and the lack of a suitable technique to study them, because, less than 1% of soil microbes can be cultured in laboratory (Bardgett, 2005). Development of phospholipids fatty acid (PLFA) profiling and rDNA finger printing methods enables to quantify diversity and to measure composition and abundance of soil microbial communities (Tunlid and White, 1992) and has proved to be of great value in describing structure and function of soil microbial community.

Semi-natural grassland in Shiriyazaki area of northern Honshu island, Japan have been maintained without human managements for over 300 years under horse grazing (Sugiyama et al., 2008). An improved grassland is located 1 km away from the semi-natural grasslands. The objective of the present chapter is to discuss about the effects of management activities on the structural changes of plant and soil microbial communities and relationships between plant and soil microbial community structures.

2. MATERIALS AND METHODS

2.1. Study Site Description

The study was conducted in semi-natural and improved grasslands of Shiriyazaki (41°25′N and 141°26′E) of Aomori prefecture located in the northern Honshu island, Japan (Figure 1). In this area, annual mean temperature is 9.2°C, and the total annual precipitation is 1298 mm. Shiriyazaki has developed about 5-km^2 area of semi-natural grassland along shoreline of the Pacific Ocean as a result of a long history of horse grazing for over 300 years.

During this time no human management activities such as fertilization and sowing of pasture species have been applied. The semi-natural grasslands of Shiriyazaki consist of two floristically distinct vegetations, short-grass and tall-grass vegetations. The short-grass vegetation is characterized by the dominance of C_4 grass, *Zoysia japonica* and low height of vegetation (6.6 cm) resulting from frequent grazing by horses, while the tall-grass vegetation, which is subject to less frequent grazing, is characterized by the dominance of C_4 tall grass *Miscanthus sinensis,* C_3 grass *Bromus japonicus* and forbs. Improved grassland, located 1 km away from semi-natural grassland, was established about 30 years ago by sowing of herbage species and has been subjected to fertilization and grazing by cows.

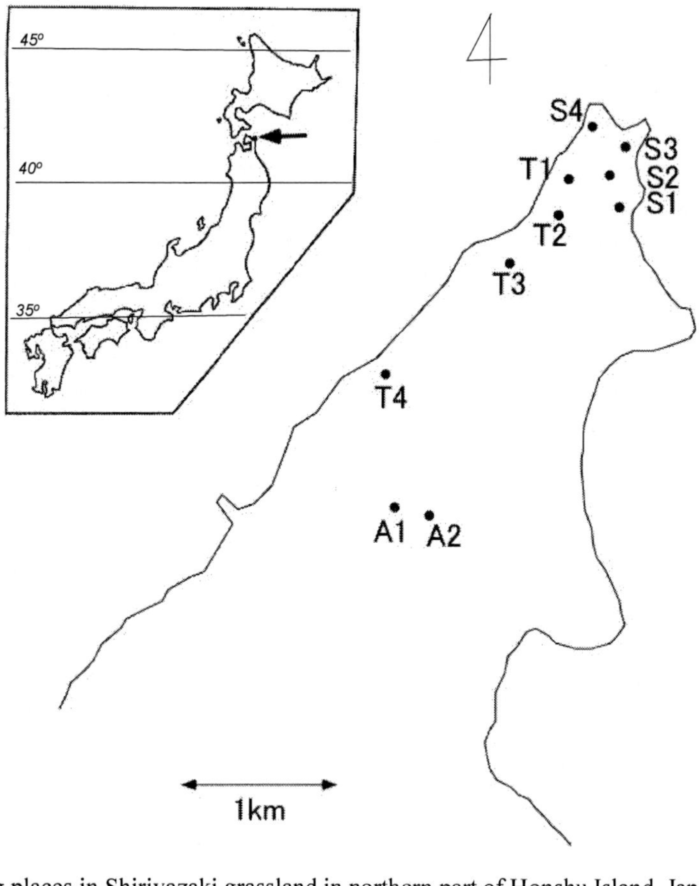

Figure 1.Sampling places in Shiriyazaki grassland in northern part of Honshu Island, Japan. S1– S4, T1–T4 and A1–A2 respectively represent sampling places of the short-grass vegetation in the semi-natural grassland, of the tall-grass vegetation in the semi-natural grassland, and of the artificial grassland. Vegetation, soil fertility and soil microbial communities were assessed in three locations of each place.

2.2. Soil Sampling and Vegetation Analysis

Soil sampling and vegetation measurements were made in September 2002. A 50 m long transect was drawn in randomly chosen 4 sites from each of the short-type and the tall-type grasslands and 2 sites from improved grasslands (Figure 1). Soil sampling and vegetation measurements were made on the three positions along the transect, 5 m, 25 m and 45 m from the starting point of each transect. Thus, measured locations were 12 for each of the short-grass and the tall-grass and 6 for the improved grassland. A plastic frame of 1 m x 1 m that is subdivided into four quadrates of 0.5 x 0.5 m was placed in each location of the transect. Plant height was recorded and component species were identified and their relative coverage was recorded in each of four quadrates.

Soil samples (40 x 40 x 60 mm deep) were then collected from the center of each quadrate and were put into a plastic bag. After moving the soil to the laboratory, the soil was sifted through a 2-mm sieve to remove large rocks, roots, and macrofauna. The soils were then kept at 4°C for further analysis.

2.3. Soil Physico-chemical Properties

Soil nitrogen content was measured before and after incubation at 25°C for 30 days. Nitrogen was extracted from 5 g fresh soil in 80 ml 2N KCl. The samples were then shaken for 1h, and filtered through a filter paper. Amount of available N (NO_3^- and NH_4^+) was determined by Kjeldahl method and the content was expressed as mg N $100g^{-1}$ dry soil. Nitrogen content measured after 30 days of incubation was used as available N.

Soil P content was determined by colorimetric method after extraction from 1 g dry soil in 0.002N H_2SO_4 solution and expressed as μg P_2O_5 g^{-1}dry soil. Gravimetric soil moisture was determined by weight loss after drying 10g soil at 60°C for 24h. Except microbial analyses, all soil properties were determined in quadruple replicates for each location.

2.4. PLFA Analysis

PLFA analysis was performed by following a previously described method (Hossain et al., 2007), which is a modified procedure of that explained by other study (Frostegård et al., 1993). Lipids were extracted from fresh soils equivalent to 7.0 g dry weight using a composite mixture of 4 replicates for each of the 30 locations. PLFA was analyzed using a gas chromatograph (GC-14B; Shimadzu Corp., Kyoto, Japan) that was equipped with a capillary column of 30 m with 0.25 m inner diameter (SPB-1; Supelco, Bellefonte, USA). In addition, PLFA mixtures extracted from representative soils were analyzed using GC-MS (PQ2010; Shimadzu Corp., Kyoto, Japan) to identify each PLFA peak. We found a total of 40 PLFA peaks and 33 of them were identified. For each sample, the abundance of individual fatty acid methyl-esters was calculated from the comparison with internal standard. Total PLFA was expressed as μg PLFA g^{-1} dry soil and the abundance of microbial groups were expressed as %nmol PLFA g^{-1} dry soil.

The fatty acids, i15:0, a15:0, 15:0, i16:0, 17:0, i17:0, cy17:0, 18:1ω7, and cy19:0 were used for bacteria biomarker (Frostegård and Bååth, 1996; Fostegard et al,. 1993; Griffiths et al., 1999); i-14:0, i-15:0, a-15:0, i-16:0, i-17:0, and a-17:0 for Gram-positive bacteria (Griffiths et al., 1999); cy-17:0, cy-19:0, and 18:1 ω 7c for Gram-negative bacteria (Porazinska et al., 2003), 18:2 ω 6 for fungi (Federle, 1986), 16:1 ω 5c for mychorrhizae and 10Me18:0 for actinomycetes (Bossio and Scow, 1998; Zak et al., 2003). In addition, monounsaturated, saturated, and hydroxylated PLFA biomarkers were classified as independent microbial groups based on the chemical structure of the fatty acid methyl esters.

2.5. DNA Extraction, PCR, and DGGE Analysis

We performed DNA extraction using a kit (Ultra Clean Soil DNA kit; Mo Bio Laboratories, Solana Beach, CA, USA). 1 g soil was processed according to the protocol. We measured the prokaryote community by amplifying 16S rDNA fragments using the primers of 341F-GC (CGC CCG CCG CGC GCGGCG GGC GGG GCG GGG GCA CGG GGG GCC TAC GGGAGG CAG CAG) and 534R (ATT ACC GCG GCT GCT GG). For the eukaryote community, PCR was performed using universal internal transcribed spacer (ITS) primers of rDNA fragments (200–450 bp): ITS1 (5′-TCC GTA GGT GAA CCT GCG G-3′) and ITS2 (5′-GCTGCG TTC TTC ATC GAT GC-3′). The reaction mixture (25 μL) consisted of 5 ng template DNA, 2.5 μL of 10 × PCR buffer (100 mol Tris-HCl, 500 mol KCl, 15 mol $MgCl_2$), 2 μL of deoxyribonucleotide triphosphate mixture (2.5 mol of each dNTP), 0.5 μM of each primer, and 0.3U of TaKaRa Taq (Takara Bio, Otsu, Japan). After 5 min of denaturation at 94°C, 35 thermal cycles of 30s at 94°C, 45s at 58.4°C and 3min at 72°C were performed, followed by an extension step at 72°C for 10min. Then, denaturing gradient gel electrophoresis (DGGE) analyses were performed using a universal mutation detection system (Dcode; Bio-Rad, Richmond, CA, USA) with a 6% acrylamide gel containing a gradient of 30–65% denaturant (100% denaturant contains 7 mol urea and 40% formamide). We applied 24 μL of PCR samples to the DGGE gel; then DGGE was performed in 1 × TAE Buffer (40 mol Tris/acetate, pH 8; 1 mol ethylene diamine tetra acetic acid, EDTA) at 60°C at a constant voltage of 100 V for 12 h. After staining (SYBR Green 1; Cambrex Bio Science, Rockland, ME, USA), the DGGE gels were recorded into digital images and the DNA band number was analyzed using image-processing software (IMAGE-J; National Institutes of Health, Bethesda, MD, USA) after subtracting background noise.

2.6. Statistical Analysis

One-way analysis of variance and multiple comparison tests (Tukey-Kramer HSD) were used to determine statistical difference between the vegetation types. Linear correlation analysis was performed to examine the relationships among fatty acid composition, soil physico-chemical properties, and plant functional type composition. Data analysis was conducted in JMP 4.0 (SAS Institute, Cary, NC, USA).

3. Results

3.1. Plant Vegetation Structure

Differences in plant diversity, functional type composition, and dominant species among the semi-natural tall-type and short-type and the improved grasslands are shown in Table 1. Plant species richness showed a significant difference among the three vegetations (F = 33.17,

$P < 0.0001$). Number of species in each quadrate (0.25 m^{-2}) ranged from 4-14, 7-18, and 2-7 in short-grass, tall-grass, and improved grasslands, respectively (Table 1). Mean species richness in the tall-grass (12.33) is about three times higher than that in the improved grassland (4.21) and the short-grass was intermediate (7.83). Shannon's diversity index also showed similar trend of variation ($F = 38.61$, $P < 0.0001$, data not shown) among the three vegetations. The diversity index in the tall-grass community (2.07) was more than double than that in the improved grassland (0.89), and the short-grass community (1.20) was intermediate.

The three vegetations also differed significantly in the composition of plant functional types; C_4 grass ($F = 56.6$, $P < 0.0001$), C_3 grass ($F = 61.24$, $P < 0.0001$), legumes ($F = 10.22$, $P < 0.0006$) and forbs ($F = 56.46$, $P < 0.0001$). In the short-grass community, more than 60% of coverage per quadrate belonged to C_4 grass and followed by forbs, C_3 grass and legumes. The tall-grass community was dominated by the forbs (about 50%) and followed by C_3 grass, C_4 grass, and legumes. About 80% of the coverage per quadrate in the improved grassland belonged to C_3 grass and the rests were to legume and forbs. Legumes were significantly higher in the improved grassland than the short-grass and the tall-grass vegetations. The improved grassland did not contain any C_4 grasses.

The dominant species for the short-grass were *Zoysia japonica, Hydrocotyle sibthorpioides*, *Poa pratense, Geum japonicum*, and *Trifolium repens*, while *Bromus japonicus*, *Miscanthus sinensis*, *Cirsium oligophylum, Zoysia japonica, and Clematis ptarmiflora,* for the tall-grass, and *Lolium perene*, *Dactylis glomerata*, *Phleum pratense*, *Poa pratense* and *Trifolium repens* for the improved grassland.

Average plant height differed significantly among the three vegetations ($F = 86.14$, $P < 0.0001$). The vegetation heights of the tall-grass (27.19 cm) and of the improved grassland (24.05 cm) were about four times higher than that of the short-grass (6.61 cm) vegetation.

3.2. Soil Physico-chemical Properties

Available N and P contents significantly differed among the three grassland soils. Available N content in the improved grassland (10.78 mg N 100g^{-1} dry soil) was more than double than that in both short-grass (4.82 mgN 100 g^{-1} dry soil) and tall-grass (4.50 mgN 100 g^{-1} dry soil) soils (Table 2). The mean soil phosphorus content was also 4 to 5 times higher in the improved grassland (234.47 μg/ g dry soil) than that in the short-grass (50.06 μg g^{-1} dry soil) and tall-grass (41.86 μg g^{-1} dry soil) soils (Table 2). Soil pH ($F = 0.662$, $P = 0.55$, data not shown) and moisture content ($F = 0.11$, $P = 0.892$) did not vary significantly among the three soil types (data not shown).

Table 2. Physico-chemical properties of the soils of seminatural short-type and tall-type and the improved grasslands in Shiriyazaki located in the northern Honshu island, Japan.

	Short-type	Tall-type	Improved	*P*-value
Available N	4.82 ± 0.67^b	4.50 ± 0.53^b	10.78 ± 0.64^a	<0.0001
Phosphorus	50.06 ± 3.38^b	41.86 ± 2.07^b	234.47 ± 47.30^a	<0.0001

Table 1. Plant diversity, functional type composition and dominant species in the seminatural tall- type and short-type and the improved grasslands in Shiriyazaki located in the northern Honshu island, Japan.

		Shortgrass	Tallgrass	Improved	*P*- value
Plant diversity	Species richness	7.83 ± 0.705^b	12.33 ± 0.59^a	4.21 ± 0.42^c	<0.0001
	Shannon *H*	1.2 ± 0.09^b	2.08 ± 0.06^a	0.89 ± 0.19^b	<0.0001
Functional type	C4	61.64 ± 5.71^a	11.72 ± 2.45^b	0 ± 0^c	<0.0001
	C3	15.67 ± 4.52^b	37.35 ± 2.96^a	11.76 ± 4.80^c	<0.0001
	Legumes	4.11 ± 0.98^b	1.37 ± 0.42^c	8.67 ± 2.11^a	0.0005
	Forbs	20.24 ± 2.28^b	48.74 ± 2.92^a	6.62 ± 2.88^c	<0.0001
Dominant species (%)	C3	*Poa pratense (5.8)*	*B. japonicus (15.7)*	*Lolium perene (43.9)*	
		Festuca ovina (2.9)	*F. ovina (4.2)*	*Dactylis glomerata (10.9)*	
		Bromus japonicus (2.6)		*Phleum pratense (10.0)*	
				Poa pratense (7.6)	
	C4	*Zoysia japonica (45.7)*	*Miscanthus sinensis (9.1)*	-	
			Z. japonica (4.2)		
	Forbs	*Hydrocotyle sibthorpioides (6.2)*	*Cirsium oligophyllum (4.8)*	-	
		Geum japonicum (2.3)	*Clematis termiflora (3.5)*		
		Geranium erienthum (2.2)	*G. japonicum (2.)*		
		Plantago asiatica (2.2)	*Potentilla fragarioides (2.0)*		
		Plantago lanceolata (1.9)	*Oxalis foutana (1.5)* *Achillea ptarmica (1.2)*		
	Legumes	*Trifolium repens (2.8)*	-	*T. repens (5.3)*	

´-´ indicates absence of species

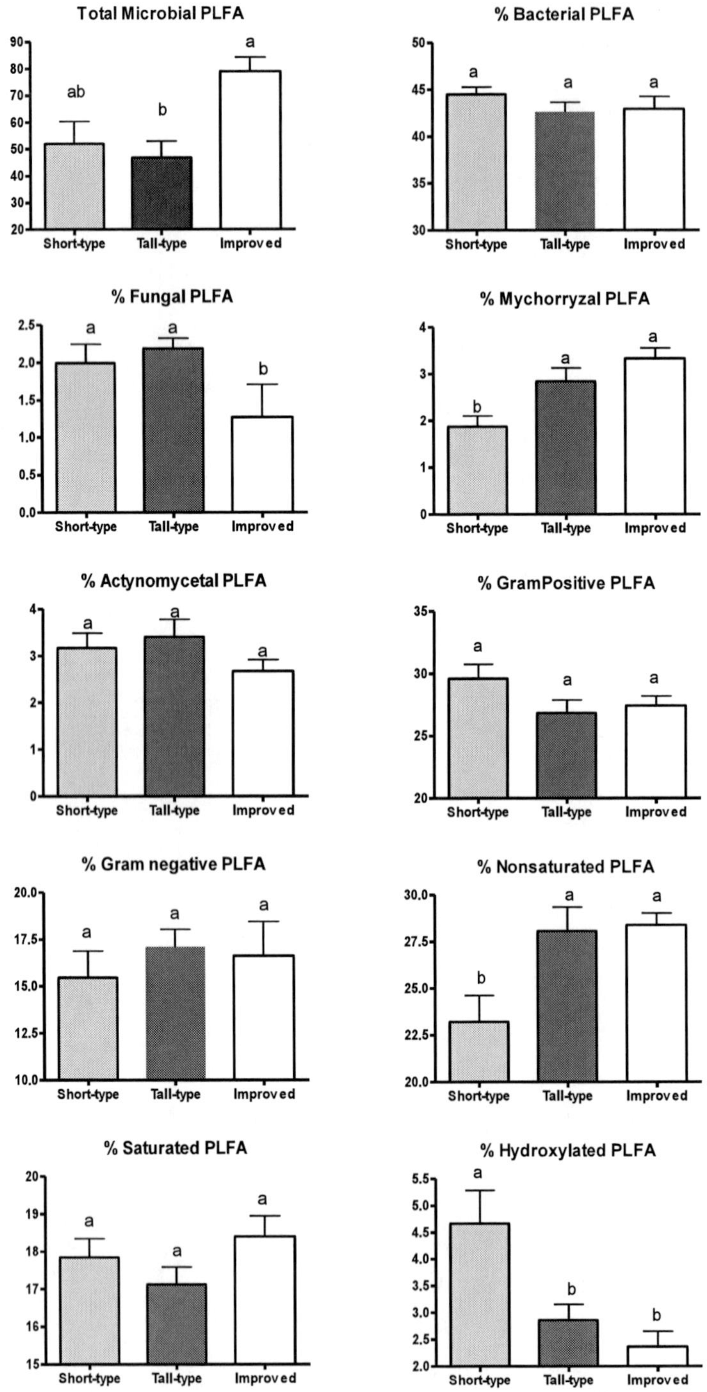

Figure 2. Microbial community structure in soils of the seminatural short-type and tall-type and improved grasslands in Shiriyazaki located in the northern Honshu island, Japan. Total microbial biomass and the relative abundance of microbial groups are expressed as μgPLFA g^{-1} dry soil and %nmol PLFA g^{-1} dry soil, respectively. Different letters indicate significant difference at $P < 0.05$.

3.3. Soil Microbial Community Structure-PLFA Data

Soil microbial community structure of the three grasslands is shown in Figure 2. Microbial biomass as determined by total PLFA mass showed significant difference (F = 3.88, P = 0.03) among the three grasslands. The improved grassland showed the highest amount of biomass (78.97 μg PLFAg^{-1} dry soil), followed by the short-grass (51.96 μg PLFA g^{-1} dry soil) and the tall-grass (46.77 μg PLFA g^{-1} dry soil).

The proportion of bacteria, including Gram-positive and Gram-negative, did not differ significantly among the grasslands, while the proportion of fungal biomass significantly differed. The tall-type and the short-type had significantly higher proportion of fungal abundance than the improved grassland. However, mycorrhizal proportion was significantly lower for the short-grass than that for the tall-grass and the improved grassland. The proportion of actinomycetal and saturated PLFA showed no significant difference among the three grasslands. The short-grass showed significantly higher proportion of monounsaturated PLFA mass and significantly lower hydroxylated PLFA mass than the tall-type and the improved grassland. PLFA peak number was significantly (P = 0.032) higher in tall-grass (34.92) than the improved grassland (31.50) and the short-grass was intermediate (33.75).

3.4. Soil Microbial Community Structure-rDNA Data

As shown in Table 3, although the three grassland types did not differ significantly in bacterial DNA band number, they differed significantly in fungal DNA band number; tall-grass showed the significantly (P = 0.034) higher number (39.08) than the improved grassland (22.33) and the short grass was intermediate (33.00).

Table 3. Microbial diversity as determined by counting DNA band number after PCR-DGGE analysis and peak number of phospholipid fatty acids.

	Short-type	Tall-type	Improved	P-value
Bacterial DNA band number	32.42±2.18a	33.83±2.13a	28.50±3.02a	0.37
Fungal DNA band number	33.00±3.46ab	39.08±3.21a	22.33±4.54b	0.024
PLFA peak number	33.75±0.509ab	34.92±0.95a	31.50±0.50b	0.032

3.5. Linkage between Plant Functional Types and Soil Microbial Communities

Table 4 shows the coefficients of correlations of plant functional types and soil fertility with relative abundance of each microbial group. Total microbial biomass showed a significant positive correlation with the proportion of legumes (r = 0.457, $P < 0.05$) and a negative one with the proportion of the forb (r = -0.369, $P < 0.05$). The proportion of legume also showed a negative correlation with the relative abundance of fungi (r = -0.480, $P < 0.01$). The relative abundance of AM fungi positively correlated with the proportion of C_3 grasses and negatively with C_4 grasses (r = -0.460, $P < 0.05$). C_3 grass also showed a negative correlation with the relative abundance of hydroxylated PLFA (r=-0.442, $P < 0.05$).

Table 4. Coefficients of correlation of soil total microbial biomass (μgPLFAg^{-1} soil) and the relative abundance of microbial groups (%nmol PLFAg^{-1} soil) with soil nutrients and plant functional type composition (%) (n=30).

	Soil nutrients		%Functional types			
	N	P	C4 grass	C3 grass	Legumes	Forbs
Total PLFA	0.271	0.31	-0.117	0.311	0.457 *	-0.369 *
Bacteria	-0.051	-0.221	0.098	-0.148	0.161	0.031
Fungi	-0.261	-0.116	0.213	-0.324	-0.48 **	0.245
AM Fungi	0.199	0.414	-0.46 *	0.556 **	0.068	-0.102
Actinomycetes	-0.007	-0.305	0.016	-0.206	-0.183	0.324
Gram positive Bacteria	0.136	0.039	0.126	-0.109	0.007	-0.07
Gram negative Bacteria	-0.167	-0.186	-0.088	0.034	0.096	0.098
Monounsaturated PLFA	-0.026	0.13	-0.332	0.335	0.027	0.042
Saturated PLFA	0.203	0.213	0.017	0.149	0.304	-0.3
Hydroxylated PLFA	-0.169	-0.188	0.447 *	-0.442	-0.206	-0.052

* and ** indicate significant at $P > 0.05$ and 0.01%, respectively.

In contrast to plant functional types, soil fertility did not show clear correlations with the relative abundance of microbial groups. Soil N content did not show any significant correlations with the relative abundance of each microbial group. The only significant correlation was found between soil phosphorus content and the relative abundance of AM fungi. These results suggest that plant species composition has greater effects on soil microbial community structure than does soil fertility.

4. Discussion

The results of the present study showed a significant difference in total microbial biomass, fungal biomass, and mycorrhizal biomass between semi-natural and improved grasslands. The improved grassland was dominated by productive C_3 grass and legume species, while seminatural grasslands were dominated by C_4 grass species (the short-grass) or by many forbs species (the tall-grass). The correlation analysis also showed that microbial biomass positively correlated with the proportion of legume and negatively with the proportion of forbs (Table 4). Litter quality is an important factor that determines structure and activity of soil microbes. C_3 grass and legumes are characterized by low C:N ratio and low lignin content compared to C_4 grass and forbs (Wedin and Tilman, 1990) and have been shown to stimulate microbial biomass production and microbial activity, leading to enhanced N mineralization (Stephan et al., 2000; Porazinska et al., 2003). Therefore, high N mineralization rate and high activity of microorganisms of the improved grassland seem to be related to dominance of C_3 grass and legume, which can provide high quality litter to the belowground soil biotic community. The lowest amount of microbial biomass in the semi-natural grasslands could be responsible for dominance by the forbs or C_4 grasses. Fungi are the organisms adapted to low quality litters, which contain low contents of N and P and high content of cell wall, phenol and lignin (Dix and Webster, 1995). It has been shown that forb litters promote fungi-dominated microbial community than bacteria-dominated microbial community (Wardle et al., 2003). Therefore, dominance by the forbs (about 50%) in the tall-grass community seems to be responsible for high fungal abundance.

The significantly higher diversity of fungi in the seminatural tall-type grassland than that in the improved grassland could be related with the higher heterogeneity of resource availability and microenvironments resulted from higher plant diversity (Bardgett, 2005). The result was consistent with previous results (Sugiyama et al., 2008). Although bacterial DNA band number did not show significant difference among the three grasslands it might be related with other factor like soil fertility (Sugiyama et al. 2008). This result suggest that bacterial and fungal community structure in soil is maintained by different mechanisms; bacteria is influenced by soil physical properties while fungi are by disturbance as suggested by Hossain and Sugiyama (2010).

The difference in plant species composition and soil fertility between the tall-grass and the short-grass vegetations are mainly due to the difference in grazing intensity between them. A number of studies (Bardgett et al., 2001) have shown that grazing intensity has important effects on soil biota through its influence on quantity and quality of plant litters and amount of animal waste deposition. Foliar herbivory can cause increase in carbon release into soils, which leads to relative increase in the role of bacteria in soil decomposition processes

(Bardgett et al., 2001; Hamilton and Frank, 2001). In this study, the improved grassland had higher total microbial biomass but lower fungal biomass than the tall-type grassland. The short-grass had significantly lower proportion of mycorrhizal biomass than the tall-grass. Gehring and Whitham (1994) suggested that arbuscular mycorrhizal infection of roots is restricted by foliar herbivory, which is consistent with this study. Therefore, the differences in activity and structure of soil microbial community between the tall-grass and the short-grass vegetations could be mainly caused by the difference in grazing intensity by horses.

Soil fertility is closely related to changes in plant species composition in grasslands (Tilman 1987) and thus plant species effects on soil microbial communities are easily confounded with effects of soil fertility. The three vegetations in this study also differed in soil nitrogen and phosphorus content (Table 2). The correlation analysis (Table 4) showed that plant functional type had greater effects on the relative abundance of microbial groups than did soil fertility. Bardgett et al. (1999) did not find any effects of fertilizer application on biomass and structure of microbial community in perennial grasslands. Innes et al. (2004) showed complex interactive effects of plant species and soil fertility on biomass and abundance of soil microorganisms. Therefore, activity and structure of soil microorganisms are likely to be regulated more by plant species composition than by soil fertility (Bardgett et al., 1999).

Less than 1% of bacteria can be cultured in laboratory (Bardgett, 2005). This methodological limitation has hampered the ecological study on soil microbial community over the decades. However, biochemical-based methods such as PLFA profiling and molecular-based rDNA finger printing methods enable to evaluate microbial diversity without culture technique (Kirk et al., 2004). PLFA method, which is based on the difference in chemical structure of fatty acids constituting cell membrane, can not identify lower level of microbial groups than DNA-based methods, which is based on the difference in base sequence of DNA. Particularly, PLFA method has low resolution for identification of fungi. In this study, only two groups of fungi were identified by PLFA analysis. On the other hand, PLFA analysis can measure abundance and diversity of each microbial group at the same time, while current DNA-based methods, which utilize PCR amplification of DNA segment, cannot estimate abundance of each group. Since information of abundance and diversity of biological community is essential for analysis of community structure, both PLFA profiling and rDNA finger printing will be useful tools for analyzing soil microbial community.

Over all, the present study reveals that management activities can alter the structure of plant vegfetation and soil microbial community structure. The result also indicates that changes of microbial community could be related with the changes of plant communities aboveground via grazing intensity. Thus, human management, grazing animals, plant communities, and the soil microbial communities in the Shiriyazaki grassland ecossytems are interacting on each other.

5. Acknowledgments

The authors are grateful to the Ministry of Education, Culture, Sports, Science and Technology, Japan for financial assistance in conducting this research.

REFERENCES

Bardgett RD. 2005. *The Biology of Soil: A community and Ecosystem Approach*. Oxford University Press, Oxford.

Bardgett RD, Cook R. 1998. Functional aspects of soil animal diversity in agricultural grasslands. *Applied Soil Ecology*, 10:263-276.

Bardgett RD, Mawdsley JL, Edwards S, Hobbs PJ, Rodwells JS, Davies WJ. 1999. Plant species and nitrogen effects on soil biological properties of temperate upland grasslands. *Functional Ecology*, 13:650-660.

Bardgett RD, Jones AC, Jones DL, Kemmitt SJ, Cook R, Hobbs PJ. 2001. Soil microbial community patterns related to history and intensity of grazing in sub-montane ecosystems. *Soil Biology and Biochemistry*, 33:1653-1664.

Bossio DA, Scow KM. 1998: Impacts of carbon and flooding on soil microbial communities: phospholipids fatty acid profiles and substrate utilization patterns. *Microbial Ecology*, 35:265-278.

DeDeyn GB, van der Putten WH. 2005. Linking aboveground and belowground diversity. *Trends in Ecology and Evolution*, 20:625-633.

Dix NJ, Webster J. 1995. Fungal Ecology, Chapman & Hall, London.

Frostegård Å, Bååth E. 1996. The use of phospholipids fatty acid analysis to estimate bacterial and fungal biomass in soil. *Biology and Fertility of Soils*, 22:59-65.

Frostegård Å, Bååth E, Tunlid A. 1993. Shifts in the structure of soil microbial communities in limed forests as revealed by phospholipids fatty acid analysis. *Soil Biology and Biochemistry*, 25:723-730.

Griffiths BS, Ritz K, Ebblewhite N, Dobson G. 1999. Soil microbial community structure: Effects of substrate loading rates. *Soil Biology and Biochemistry* 31:145-153.

Hamilton EW, Frank DA. 2001. Can plants stimulate soil microbes and their own nutrient supply? Evidence from a grazing tolerance grass. *Ecology* 82:2397–2402.

Innes L, Hobbs PJ, Bardgett RD. 2004. The impacts of individual plant species on rhizosphere microbial communities in soils of different fertility. *Biology and Fertility of Soils*, 40:7-13.

Hooper DU, Chapin FS, Ewel JJ, Hector A, Inchausti P, Lavorel S, Lawton JH, Lodge DM, Loreau M, Naeem S, Schmid B, Setälä H, Symstad AJ, Vandermeer J, Wardle DA. 2005. Effect of biodiversity on ecosystem functioning: A consensus of current knowledge. *Ecological Monograph*, 75:3-35.

Hossain MZ, Okubo A, Sugiyama S. 2007. Local variation in soil microbial community structure in artificial and semi-natural grasslands. *Grassland Science*, 53:165-171.

Hossain Z, Sugiyama S. 2010. Geographical structure of soil microbial communities in northern Japan: Effects of distance, land use type and soil properties. *European Journal of Soil Biology*, doi:10.1016 /j.ejsobi.2010.11.007.

Kirk JL, Beaudett LA, Hart M, Moutoglis P, Klironomos JN, Lee H, Trevors JT. 2004. Methods of studying soil microbial diversity. *Journal of Microbiological Methods*, 58:169-188.

Porazinska DL, Bardgett RD, Blaauw MB, Hunt W, Parsons AN, Seastedt TR, Wall DH. 2003. Relationships at the aboveground-belowground interface: plants, soil biota, and soil processes. *Ecological Monographs*, 73:377-395.

Schlesinger WH. 1997. *Biogeochemistry: an Analysis of Global Change*. Academic Press Inc., San Diego, CA.

Stephan A, Meyer AH, Schmid B. 2000. Plant diversity affects culturable soil bacteria in experimental grassland communities. *Journal of Ecology*, 88:988-999.

Sugiyama S, Hossain MZ, Okubo A. 2008. Relationships between soil microbial diversity and plant community structure in seminatural grasslands. *Grassland Science,* 54:117–124.

Tilman D. 1987. Secondary succession and the pattern of plant dominance along experimental nitrogen gradients. *Ecological Monographs*, 57:189-214.

Tunlid A, White DC. 1992. Biochemical analysis of biomass, community structure, nutritional status, and metabolic activity of microbial communities in soil. *Soil Biochemistry* (eds G. Stotzky & J.M. Bollag), Vol. 7, pp. 229-262. Marcel Dekker, New York, NY.

Wardle DA. 2002. Communities and ecosystems: Linking the aboveground and belowground components. Princeton University Press. Princeton, NJ, USA.

Wardle DA, Yeates GW, Williamson W, Bonner KI. 2003. The response of a three trophic level soil food web to the identity and diversity of plant species and functional groups. *Oikos*, 102:45-56.

Wedin DA, Tilman D. 1990. Species effects on nitrogen cycling: a test with perennial grasses. *Oecologia*, 84:433-441.

Zak DR, Holmes WE, White DC, Peacock AD, Tilman D. 2003. Plant Diversity, soil microbial communities, and ecosystem function: are there any links? *Ecology*, 84:2042-2050.

In: Grasslands: Types, Biodiversity and Impacts
Editor: Wen-Jun Zhang

ISBN: 978-1-61470-555-0

Chapter 6

CONSERVATION AND MANAGEMENT OF ALKALI GRASSLAND BIODIVERSITY IN CENTRAL-EUROPE

Török Péter[1*], Kapocsi István[2] and Deák Balázs[2]
[1] University of Debrecen, Department of Ecology
H-4032 Debrecen, Egyetem tér, Hungary.
[2] Hortobágy National Park Directorate, Hungary

ABSTRACT

Grasslands are vital landscape elements in Europe. Recently, the 180 million hectares of grasslands have a crucial role in maintaining the landscape level biodiversity. Alkali grasslands are typical in Central- and Eastern part of Europe, with large areas in the Carpathian-basin. These types of grasslands were not the most favorable targets of arable farming, but large areas affected by mineral fertilization, drainage, soil melioration and/or commercial seeding in the last 60 years. In our paper we present important vegetation characteristics, species composition and management of five different grassland types from the open annual alkali pioneer swards to tall grasses dominated wet alkali meadows. In general, alkali grasslands are usually species poor communities characterized by short (*Festuca pseudovina*, *F. rupicola*, *Poa angustifolia*) or tall grasses (*Alopecurus pratensis, Elymus repens*). They harbor several steppe endemics (e.g. *Plantago schwarzenbergiana*, *Cirsium brachycephalum, Limonium gmelinii* ssp. *hungarica*, *Puccinellia limosa* and *P. peisonis*) and halophyte species (*Salicornia prostrata, Salsola soda, Suaeda pannonica, S. maritima,*), adapted to high salt contents of soil. According to the uneven pattern of soil salt and water alkali grasslands are spatially very diverse. Maintaining alkali grasslands the extensive grazing mostly by cattle and sheep is essential. Nowadays, in large areas of alkali grasslands former grazing are ceased or replaced with mowing. This resulted in a change of species composition, decreased richness and/or litter accumulation. Alkali grasslands are refugees of alkali steppe vegetation, thus, restoration and preservation of their biodiversity have a high conservation priority in Habitats Directive of the EU (Pannonic salt steppes and salt marshes, 1530).

Keywords: alkali grassland types; salt; grazing; mowing; grassland recovery.

1. Introduction

Conservation of grassland biodiversity is of top priority in conservation policy of Europe and elsewhere (Nösberger and Rodriguez 1996). Grasslands are vital elements of traditional landscape and of crucial importance for biodiversity conservation (Wallis De Vries et al. 2002). Grasslands also provide important ecosystem services for recreation, sport and tourism (Isselstein et al. 2005). Natural grasslands are in constant decline in many parts of Europe. This considerable decline is mostly caused by the conversion of grasslands into croplands and tree plantations (Hald and Vinther 2000, Poschlod and WallisDeVries 2002). In western part of Europe this process was very intensive in countries with improved agriculture (Kiehl et al. 2010). In the remained grasslands ceased or altered traditional management by grazing and mowing (Valkó et al. 2011), agricultural intensification in a form of improved use of fertilizers, pesticides, and commercial seeding resulted in a rapid decline of former species richness (Bakker and Berendse 1999). Thus, the conservation and restoration of grasslands became a high priority action worldwide.

In Eastern and Central Europe in the former socialist era many grasslands were ploughed up, and the use of fertilizers was also increased; however extent grasslands in areas with low potential of crop-production were maintained by low intensity farming. After the collapse of the state own agricultural cooperatives the socio-economical changes associated with cheap import and insufficient funds of the new land-owners were resulted in a decrease of 50-70% of former livestock numbers in this region and a large scale abandonment of semi natural pastures and mown grasslands (Isselstein et al. 2005). The total extension of grasslands was decreased by 14% caused mainly by ceased former management between 1989 and 2007 in Hungary.

Among the remnant grassland fragments of near natural state, alkali grasslands are one of the best preserved grassland habitats in Europe. These types of grasslands are utilized as pastures, because their poor soil quality and fluctuating water regime unsuitable for intensive agriculture and forestry. Due to these facts, their unique flora and fauna is well preserved till now. In Habitats Directive which is a cornerstone of Europe's nature conservation policy alkali habitats have an important role. These types of habitats of Carpathian basin were included into Natura 2000 network as “Pannonic salt steppes and salt marshes (1530)” registered as priority habitat type of the EU. In the present paper we aimed at to discuss the most important alkali grassland types, their traditional management, current conservation status, and restoration possibilities.

2. Soil Conditions and Climate

Continental alkali grasslands are influenced by the high soil salt content and dynamic changes of water regime. These types of grasslands are generally located in soils with high levels of groundwater and at least moderate salt concentration (Na_2CO_3, NaCl, rarely K_2CO_3 and KCl) and a high pH (8-10) under a continental climate. The soil salt content is originated from eroded debris transported by groundwater from volcanic mountains and sodium, calcium and magnesium content of the bedrock (predominantly loess). Due to the intensive evaporation caused by often extreme, dry and warm summer periods, the groundwater which is near to the soil surface and have high salt content transports the salt to the upper soil layers where its accumulates.

Types of alkali soils and grasslands are generally determined by the vertical position of salt accumulation zone and groundwater level. The least productive types are where the groundwater level is high and the salt accumulation zone is at or near to the soil surface. Characteristic alkali soil types are solonchak and various types of solonetz. *Solonchak* soils frequently have a considerable sand soil fraction and are characterized by high mean groundwater levels where the salt accumulation zone is near located to the soil surface. The humus content is generally low. There is generally no definite geometric soil structure. *Solonetz* soils are well structured and can be characterized with deeper groundwater levels and salt accumulation zones than solonchak soils. The steppic types are characterized by high humus content. The upper soil layer can be very hard (if dry) or soapy (if wet).

Alkali soils can develop by natural or a human induced way (Molnár and Borhidi 2003). Naturally developed alkali soils are present in the Carpathian-basin since the Pleistocene with considerable extent (Bárczi et al. 2006). Secondary alkali soils are formed after drainage or river regulation works, which was typical from the 19th century (river regulations) till the end of the socialist era (irrigation in marshes and fens) in the Carpathian basin. The lowered water levels and desiccation of habitats induced secondary salt accumulation in higher soil layers especially in former wet non-alkali meadows, marshes or even in fens (Molnár and Borhidi 2003).

3. Grassland Types

3.1. General Characteristics

Alkali grasslands are typical in regions with at least moderately continental climate, extent stands can be found from Middle-Asia (e.g. Siberia and Mongolia) to Middle- and Eastern-Europe (Ukraine, Romania, Hungary), but very sporadic in Western Europe. The largest native occurrence of alkali grasslands to the west is located in the Carpathian-basin (mostly in Hungary and Romania, but some minor grassland patches can be found in Austria and in the Czech Republic).

The dominant species of most alkali grassland types are widely distributed grass species with a wide range of humidity and salt tolerance (*Agrostis stolonifera*, *Alopecurus pratensis*, *Beckmannia eruciformis, Elymus repens, Festuca pseudovina* and *F. rupicola*). Alkali grasslands harbor several grassland species characteristic to Eurasian continental steppes and several endemics to the Carpathian basin. Several alkali grassland types harbor the same common grassland species; thus their classification is based not only on differential species and species composition but on the relative proportion of dominant species. On habitats with extremely high salt contents and rapid changes in water availability sparse halophytes dominated vegetation is typical. These types of grasslands are classified in "Annual alkali pioneer swards of steppes and lakes" when several short lived herbs are characteristic, and in "Dense and tall *Puccinellia* swards" if *Puccinellia* spp. are present with a relatively high abundance. The relatively dry, short-grass alkali grassland types are "*Artemisia* alkali steppes" and "*Achillea* alkali steppes" on meadow solonetz or solonchak. Both types are characterized by the same short grasses, but the latter one generally harbors several non alkali species (species characteristic to loess grasslands). Tall grasses dominated species poor meadows on alkali soils are classified into the "Alkali meadows" type. The "Tall herb alkali meadows and meadow steppes" harbor the same tall grass species as the former type, but are generally situated on higher elevations with higher richness of forbs characteristic for loess grasslands, alkali steppes and alkali meadows. In the followings we present these types of alkali grasslands in the Carpathian basin, in accordance with the classification of Molnár et al. (2008).

Extent homogeneous stands of a single alkali grassland type can be rarely found; various type of grasslands form generally a heterogeneous mosaic structure in accordance with the uneven pattern of soil salt contents, relief and water availability.

In a landscape characterized by alkali grasslands near to the highest elevated plateaus and point bars with loess vegetation generally stands of *Achillea* alkali steppes are situated. Near to *Achillea* alkali steppes but at lower elevations on soils with higher salt content (solonetz or solonchak) typical *Artemisia* alkali steppe vegetation can be found (Molnár and Botta-Dukát 1998; Török et al. 2011a). At the lowest elevations alkali meadows, and in the deepest depressions alkali marshes are situated.

Figure 1. Open pioneer alkali grassland with *Salicornia prostrate.*

3.2. Annual Alkali Pioneer Swards of Steppes and Lakes

These types of habitats are typical on salt lake beds and several micro-formations of alkali landscape characterized by extremely high salt accumulation at or near to the soil surface, low soil fertility and high groundwater level even in summertime. Annual alkali pioneer swards of steppes and lakes have a low vegetation cover and harbor a high number of endemic (*Puccinellia peisonis, Suaeda pannonica, S. salinaria*) and common halophyte species (*Camphorosma annua, Lepidium cartilagineum, Matricaria recutita, Plantago maritima, Salicornia prostrata, Spergularia maritima, Suaeda maritima* Fig. 1.).

3.3. Dense and Tall *Puccinellia* Swards

They are generally wet and species poor alkali habitats with high concentration of accumulated salt in the upper soil layers. Their stands are covered with shallow water in the spring. The vegetation cover usually exceeds 50% in springtime, but in dry years and/or seasons this cover can decrease to a few percent. Stands are more typically located in solonchak soils where the extension of continuous stands reaches several hectares. Most characteristic are the endemic *Puccinellia* species (*Puccinellia limosa, P. pannonica, P. peisonis*), and the commonly widespread *P. distans,* however the vegetation composition changes very dynamically in time. In wet years several *Juncus* species and *Alopecurus pratensis* can reach considerable cover. In dry years short-lived forbs like *Artemisia santonicum, C. annua, Cerastium dubium, M. recutita, Pholiurus pannonicus, Plantago. maritima* and *P. tenuiflora* become frequent. *Puccinellia* swards are generally used as pastures grazed by cattle.

3.4. *Artemisia* Alkali Steppes

Artemisia alkali steppes are characterized by short grasses (mainly *F. pseudovina*) and two subspecies of *A. santonicum* (*A. santonicum* ssp. *santonicum* and *patens;* latter is an endemic subspecies) regularly used as pastures. These grassland types are generally covered by shallow water (a few mm) in springtime. In the dry summer polygonal soil splits may appear on the surface. *Bupleurum tenuissimum, C. dubium, Gypsophila muralis, Limonium gmelinii* ssp. *hungarica, Lotus tenuis, Podospermum canum, Ranunculus pedatus* and several *Trifolium* species. are frequent species of alkali steppes. Salt tolerant species like *Aster tripolium ssp. pannonicum, Atriplex littoralis, L. cartilagineum, P. maritima, Suaeda* spp., are widespread.

Artemisia alkali steppes have several endemic species like *A. tripolium* ssp. *pannonicum, L. gmelinii* ssp. *hungarica, Suaeda* spp., *Puccinellia* spp. and *Plantago schwarzenbergiana* which is present in disturbed stands. If suitable water conditions are met (e.g. in springtime) and overgrazing occur several short-lived species (*Erophila verna, Hordeum hystrix, M. recutita* and *Polygonum aviculare*) are present. In wet years in open patches several species characteristic to annual alkali pioneer swards (*Myosurus minimus, P. limosa, Scleranthus anuus*) can be found.

3.5. *Achillea* Alkali Steppes on Meadow Solonetz

Achillea alkali steppes are closed, short-grass steppes on solonetz soils with a moderate salt contents (meadow and chernozem solonetz), characterized by short grasses (*F. pseudovina, F. rupicola, Koeleria cristata, Poa angustifolia* and in heavily grazed stands *Cynodon dactylon*, Fig. 2). Considerable stands of *Achillea* steppes were originated due to secondary salt accumulation. Several species characteristic also for loess grasslands are common (*Dianthus pontederae, Filipendula vulgaris, Silene viscosa, Salvia austriaca*). Because of the relatively low soil salt concentration only a few salt specialist can be found in these grasslands. *Achillea* steppes are rich in forb species, the most frequent species are

Achillea collina, A. setacea, B. tenuissimum, C. dubium, G. muralis, Inula britannica, Plantago lanceolata, P. canum, Ornithogalum kochii. Several *Trifolium* species are also present (*Trifolium angulatum, T. fragiferum, T. retusum, T. striatum, T. strictum*). These types of grasslands are traditionally used as pastures grazed by cattle or sheep. In overgrazed stands grazing tolerant species can appear (*Carduus nutans, C. dactylon, Eryngium campestre, Ononis spinosa*).

3.6. Alkali Meadows

Alkali meadows are tall grasslands dominated by tall grasses like *A. pratensis, A. stolonifera, B. eruciformis, E. repens* and *Glyceria fluitans*. This type of alkali grassland is generally covered by shallow water from early spring till midsummer. There are usually two herb layers. In the upper herb layer with the tall grasses several tall forbs are characteristic (e.g. *Lycopus europaeus*, *Lythrum salicaria, L. virgatum* and *Rumex stenophyllus*). In the second herb layer small forbs (*Cerastium dubium, Galium palustre, I. britannica, Leonurus marrubiastrum, Lotus tenuis, Lysimachia nummularia, Mentha aquatica, Mentha pulegium, Oenanthe silaifolia, Ranunculus lateriflorus, R. repens, R. sardous*) short sedges (*Carex distans, Carex melanostachya*) and rushes (*Juncus compressus, Juncus gerardii*) are present. Two endemic species *Rorippa sylvestris* ssp. *kerneri* and *Cirsium brachycephalum* are typical in most stands. These grasslands are traditionally extensively grazed or used as hay meadows.

Figure 2. *Achillea* alkali steppes in July in the Hortobágy Puszta, Hungary.

3.7. Tall Herb Alkali Meadows and Meadow Steppes

Originally these grasslands can be found in the clearings of alkali steppe oak woodlands (Bölöni et al. 2008, Molnár et al. 2008). The vegetation of these usually tall alkali grasslands are characterized by several tall grasses and herbs. In springtime they are usually wet, and get dry till late summer. Characteristic species are tall grasses as *A. pratensis* and *E. repens* and several tall forbs like *A. sedifolius*, *Rumex pseudonatronatus* and *Peucedanum officinale*. Several species of loess steppes (*F. vulgaris, Fragaria viridis, Galium verum, Peucedanum alsaticum, Stellaria graminea*) and several salt tolerant species (*Artemisia pontica, A. santonicum* and *L. gmellinii* ssp. *hungarica*) can be found. This habitat type is very rare and of high conservation value in the Carpathian basin and traditionally managed by low intensity cattle grazing.

4. Management and Threats

4.1. Traditional Management

Alkali grasslands were maintained by the large wild herbivores from the Pleistocene till the early Holocene (Bárczi et al. 2006). Parallel to the increase of the human presence in the Carpathian-basin these herbivores have disappeared. Their role was taken over by domestic grazers like cattle, sheep and horse. Nomadic herding in alkali habitats was present from the middle ages till the end of the Second World War.

Artemisia and *Achillea* alkali steppes, *Puccinellia* swards and annual alkali swards of steppes were generally used as pastures for sheep. Cattle grazing is typical on all alkali grassland types but annual pioneer swards. It is more frequent on *Achillea* and *Artemisia* alkali steppes and alkali meadows.. The wet grasslands and marshes were grazed also by Hungarian Grey cattle and Water Buffalo. The optimal stocking density was one large livestock (cattle, horse or buffalo) or 4-7 sheep per hectare. Traditionally only a negligible part of meadows were mown for producing some additional winter forage.

The most suitable time for mowing was at the beginning of flowering of the frequent grass species, usually in May. In some years the mowing time could have an overlap with the breeding time of several protected bird species, or in wet years the soil surface can be covered by water. Thus in practice, mowing of nature conservation areas is only permitted from the middle of June, but only with mowing machinery equipped with an instrument warning nesting birds (e.g. chains attached to the front of the mower).

4.2. Current Management and Threats

From the middle of the 20th century onwards the number of livestock decreased rapidly because of the economic and political changes after the Second World War. At the early years of the communist era several attempts were made to improve the poor fertility of pastures by draining water, melioration, use of fertilizers, commercial seeding and by application of pesticides. The former use by nomadic herding was in several places replaced by intensive animal farming where the livestock was kept in stables. An additional effect of the established animal framing systems was, the increased rate of linear infrastructures (e.g. roads and canals) was necessary to transport additional forage and take away animal dung.

To provide forage for stabled livestock most of the formerly grazed alkali meadows were transformed to hay meadows. The mowing by mowing machinery resulted in a more homogeneous vegetation structure than grazing and this change in land use altered the species composition of the meadows (Török et al. 2011b). The former patch dynamics were destroyed and several species adapted to grazing induced patchiness (e.g. characteristic short-lived endemics adapted to regular trampling and grazing) became rare.

It was also typical, that in several pastures livestock grazing was replaced by grazing of domestic geese. Ten thousands of domestic geese were fenced in a relatively small area which caused a dramatic decrease in the cover of grassland species and increase that of weeds in the subjected grasslands. Thus, goose grazing is not allowed in protected Natura 2000 sites, nowadays.

After large state-run agricultural co-operatives have collapsed and lands were often privatized after 1990 the cessation of former management became typical. By the cessation of management by grazing and mowing the species number, heterogeneity and number of microhabitats (formed by regular disturbance by grazing and mowing) was decreased. Abundance of frequent grassland species e.g. *A. pratensis, A. stolonifera, E. repens* in meadows, *Festuca* and *Artemisia* species in short-grasslands, and amount of accumulated litter was increased. These changes resulted in a decrease of plant species richness found also in other type of grasslands (Valkó et al. 2011). The formed homogenous grassland stands with high litter accumulation became also improper for birds (e.g. for *Charadrius alexandrinus*, *Glareola pratincola, Tringa tetanus, Vanellus vanellus*) preferring open grassland patches for nesting. The thick litter layer prohibited also the reintroduction of grazing and can increase the probability of spontaneous fires. However, it is important to stress that in lack of grazing (or mowing) no spontaneous tree and shrub invasion can be foreseen, because of the improper arboreal vegetation.

The increase of several invasive species was recently detected caused by unfavorable changes in hydrological and management status of alkali grasslands. Non-native invaders in dry alkali grasslands are *Elaeagnus angustifolia* and *Hordeum jubatum*. In abandoned wet alkali meadows and marshes the spread of the alien *Amorpha fruticosa* was detected. Three other non-native tree species *Acer negundo, Ailanthus altissima* and *Fraxinus pennsylvanica* are typical invaders in riparian forests and near to water streams. Thus, their spread to wet abandoned hay meadows with low soil salt contents can be foreseen.

5. Conservation Remarks

Several recent attempts were made to restore and conserve alkali grassland biodiversity in the Carpathian basin. In one of the largest alkali habitat complex in the Hortobágy Puszta several large scale habitat restoration projects were started from 2002 onwards, funded by the EU. To restore the historical water regime of water drained marshes and grasslands the key action was to eliminate irrigation systems (dikes and channels). In the project "Restoration of pannonic steppes, marshes of Hortobágy National Park" (LIFE02NAT/H/008634) 560 km draining channel was eliminated between 2002 and 2005 and further 400 km will be eliminated in the near future.

The largest known alkali grassland restoration in Europe was in Egyek-Pusztakócs, Hortobágy National Park, Hungary. In this project co-financed by the LIFE Nature programme and the EU (LIFE04NAT/HU/000119), altogether 665 ha of alkali grassland was restored in 2004-2008 period using low diversity seed mixtures of local provenance (for details see Török et al. 2010). To restore Important Bird Areas in alkali landscapes in the projects “Habitat management of Hortobágy eco-region for bird protection” (LIFE02 NAT/H/008638), “Sodic lake habitat restoration” (LIFENAT07 /H/000324), and “Conservation of *Otis tarda* in Hungary” (LIFE04 NAT/HU/000109) restoration of alkali steppe and wetland habitats were planned to provide feeding and breeding habitats for endangered bird species (e.g. *Chlidonias leucopterus*, *Grus grus*, *Otis tarda*). These aims are fulfilled with complex habitat restoration measures including rewetting actions, grassland recovery by sowing of low diversity seed mixtures (over 350 ha in the Kiskunsági National Park and in the Hortobágyi National Park) and reintroduction of former management by extensive grazing. Due to these efforts the nesting populations of the protected bird species were increased significantly (e.g. these increase was more than 16% for *Otis tarda* populations).

ACKNOWLEDGMENTS

Advices of colleagues in University of Debrecen (Béla Tóthmérész, Orsolya Valkó); in Hortobágy National Park (Attila Molnár, Szilvia Gőri, Sándor Konyhás, Róbert Kiss, Lajos Gál) and in Kiskunsági National Park (Ferenc Sipos, András Bankovics) during manuscript preparation is greatly acknowledged. The authors were supported by TÁMOP 4.2.1./B-09/1/KONV-2010-0007 project. The TÁMOP project is implemented through the New Hungary Development Plan, co-financed by the European Social Fund and the European Regional Development Fund.

REFERENCES

Bakker JP, Berendse F 1999. Constraints in the restoration of ecological diversity in grassland and heathland communities. *Trends in Ecology and Evolution*, 14: 63-68.

Bárczi A, Tóth TM, Csanádi A, Sümegi P, Czinkota I 2006. Reconstruction of the paleo-environment and soil evolution of the Csípő-halom kurgan, Hungary. *Quaternary International*, 156-157: 49-59.

Hald AB, Vinther E 2000. Restoration of a species-rich fen-meadow after abandonment: response of 64 plant species to management. *Applied Vegetation Science*, 3: 15-24.

Isselstein J, Jeangros B, Pavlů V 2005. Agronomic aspects of biodiversity targeted management of temperate grasslands in Europe- A review. *Agronomy Research*, 3: 139-151.

Kiehl K, Kirmer A, Donath TW, Rasran L, Hölzel N 2010. Species introduction in restoration projects – Evaluation of different techniques for the establishment of semi-natural grasslands in Central and Northwestern Europe. *Basic and Applied Ecology*, 11: 285-299.

Molnár Zs, Borhidi A 2003. Hungarian alkali vegetation: Origins, landscape history, syntaxonomy, conservation. *Phytocoenologia*, 33: 377-408.

Molnár Zs, Botta-Dukát Z 1998. Improved space-for-time substitution for hypothesis generation: secondary grasslands with documented site history in SE-Hungary. *Phytocoenologia*, 28: 1-29.

Molnár Zs, Bíró M, Bölöni J, Horváth F 2008. Distribution of the (semi-) natural habitats in Hungary I. Marshes and grasslands. *Acta Botanica Hungarica*, 50: 59-105.

Bölöni J, Molnár Zs, Biró M, Horváth F 2008. Distribution of the (semi-)natural habitats in Hungary II. Woodlands and shrublands. *Acta Botanica Hungarica*, 50:107–148.

Nösberger J, Rodriguez M 1996. Increasing biodiversity through management. *Grassland Science in Europe*, 1: 949-956.

Poschlod P, Wallis De Vries MF 2002. The historical and socioeconomic perspective of calcareous grasslands - lessons from the distant and recent past. *Biological Conservation*, 104: 361-376.

Török P, Deák B, Vida E, Valkó O, Lengyel Sz, Tóthmérész B 2010. Restoring grassland biodiversity: Sowing lowdiversity seed mixtures can lead to rapid favourable changes. *Biological Conservation*, 148: 806-812.

Török P, Kelemen A, Valkó O, Deák B, Lukács B, Tóthmérész B 2011a. Lucerne-dominated fields recover native grass diversity without intensive management actions. *Journal of Applied Ecology*, 48: 257-264.

Török P, Vida E, Deák B, Lengyel Sz, Tóthmérész B 2011b. Grassland restoration on former croplands in Europe: an assessment of applicability of techniques and costs. *Biodiversity & Conservation,* doi: 10.1007/s10531-011-9992-4.

Valkó O, Török P, Tóthmérész B, Matus G 2011. Restoration potential in seed banks of acidic fen and dry-mesophilous meadows: Can restoration be based on local seed banks? *Restoration Ecology*, 19: 9-15.

WallisDeVries MF, Poschlod P, Willems JH 2002. Challenges for the conservation of calcareous grasslands in northwestern Europe: integrating the requirements of flora and fauna. *Biological Conservation*, 104: 265-273.

In: Grasslands: Types, Biodiversity and Impacts
Editor: Wen-Jun Zhang

ISBN: 978-1-61470-555-0

Chapter 7

THE SOIL NEMATODES IN NATURAL AND SEMI-NATURAL GRASSLANDS AND THEIR USE AS BIOINDICATORS

Marek Renčo*

Parasitological Institute of Slovak Academy of Sciences,
Department of Plant Nematology, Hlinkova, Košice, Slovak Republic

ABSTRACT

Grasslands are the most widespread ecosystem worldwide. They are the natural habitats of a multitude of the grasses and herb species. They are of particular economic and ecological importance. Grasslands, as a part of natural ecosystem, represent the natural habitat for many soil microorganisms and animal species. An important part of the natural microfauna of grass ecosystems are nematodes, which are a complex of morphologically diverse species. According to their life strategy, the soil nematodes are divided into two groups: free living soil species and the plant parasitic species. Plant composition in grasslands plays a key role in determining soil nematode composition both in above and belowground resource-based mechanisms and in altering abiotic conditions. Nematodes can also play an essentials role in the retention of grasslands ecosystems. As primary consumers of saprophytic bacteria and fungi, nematodes make mineral nutrients available to higher plants. Parasitizing on grass roots, plant parasitic species affect the viability of grass plants, due to reduction of water and plant nutrient uptake, leading to nutrient deficiencies, while other nematodes are antagonists of organisms that negatively affect plant growth, and thus benefiting the plants. Nematodes are omnipresent, various, abundant, in a direct contact with soluble compounds in the soil water through their permeable cuticle, easily extracted and divided into trophic or ecological groups differing in their food source that have developed during their evolution. Representation of nematode trophic groups in soil, species diversity and abundance of genera or species of nematodes within their community may serve as an indicator of the environmental assessment of land ecosystem based on ecological and diversity indices.

Keywords: grassland, soil nematode functional diversity, bioindicators

* Correspondence address: email:renco@saske.sk

INTRODUCTION

Over the course of several millennia, the appearance of the landscapes around the word has been formed by human activity. In the past, mostly forested landscape has been gradually changed into a mosaic of forest, fields and grasslands up to the present form. Grasslands began to arise in connection with animal husbandry. Grubbing and burning forest and grazing in the forest have contributed to gradual expansion of different types of grasses and herbs that in long-term coexistence with extensive management created the multi species-diverse grassland communities.

At present, according to their formation, grasslands are generally divided into natural (prairie, steppe, savannas, shrub lands, tundra, mountain grasslands etc.), semi-natural and anthropogenic, created and maintained by human activity. The majority of grasslands in temperate climates are semi-natural. Although their plant communities are natural, their maintenance depends upon anthropogenic activities such as low-intensity farming, which maintains these grasslands through grazing and cutting regimes. Agriculturally improved grasslands, which dominate modern intensive agricultural landscapes, are usually poor in wild plant species due to the original diversity of plants having been destroyed by cultivation, the original wild-plant communities having been replaced by sown monocultures of cultivated varieties of grasses and clovers. In general, importance of grasslands is notably productive (production of fodder), landscape, recreational and aesthetic, but also serve as a ground water treatment plants and the prevention of flood (Galvánek and Viestová, 2006).

From the perspective of nature conservation, the most important are grasslands with the natural plant species assembling (semi-natural and natural). Those have extremely varied gene pool (gene importance), which can benefit humans in many areas. We have no idea how many potential valuable drugs or chemical substances useful to man are present in the common meadows plants or animals that are in relationships with them, whereas artificial grasslands (monoculture) but also those that have lost their value because they are no longer used negatively affect nature conservation (Galvánek and Viestová, 2006). Therefore it is important to avoid their disturbance e.g. heavy metals pollution and strive to maintain semi-grassland (mowing, grazing and fertilization). On the other hand, grasslands, as part of natural ecosystems, represent the natural habits of many soil microorganisms and animals.

Grassland species composition reflects the natural habitat and environmental conditions, and represents soil systems different from those of forests and other vegetation types, even from the same parent material. A key feature of grasslands is their high turnover of shoot and root biomass, and consequent large pool of labile organic matter at the soil surface. Ecological analysis showed that the grasslands are most distinguished by altitude and humidity regime. They are usually therefore divided into four ecological groups: mesophilic, thermophilic and xerophytic, hydrophilic and alpine vegetation. Though many taxonomic groups of fauna are living in grassland soils (Bardgett and Cook, 1998), nematodes are the most abundant and probably functionally the most important group.

In general, we can characterize nematodes as small, non-segmented worm-like organisms, inhabiting both terrestrial and aquatic (marine, freshwater) ecosystems. They belong to the phylum Nemata that includes plant, insect, animal and human parasites and predatory forms that consume microorganisms including bacteria, fungi, algae, and other nematodes (Platt, 1994). For practical reasons, the greatest attention is usually paid to parasites of humans, animals and plants (Háněl, 1999)

The group of soil nematodes can by characterised as the most numerous animals inhabiting the Earth surface from poles to the tropics, generaly about 0.5 to 1mm long. They live in water films and water-filled pore spaces of soil. According to life strategy, the soil nematodes are divided into two groups: free living soil species (many of them have beneficial role in the soil) and the plant parasitic species (they feed on plant with impact on agricultural production). However, most of soil nematodes feed on fungi, bacteria and other soil organisms, including plant pests and pathogens.

Densities of these organisms can by as high as 10 millions per square m (Yeates et al., 1997). Their function is irreplaceable in soil food web. They interact closely with other soil organisms and their activity affects primary production, decomposition, energy flows, and nutrient cycling, especially nitrogen cycle (Sohlenius, 1980; Freckman and Baldwin, 1990).

Role of Plants in the Composition of Nematode Fauna

A variety of environmental, biological and anthoropogenic factors related to resources can influence the soil nematode communities, nematode trophic structure and occurrence of particular species, such a s plant and soil organic matter, habitat (e.g. soil characteristics), physical and chemical properties of soil (moisture, temperature, aeration, texture, pH), climatic and geographical factors (altitude, elevation, slope), presence of other living organisms such as fungi, bacteria, viruses and other nematodes, human activities (tillage, fertilization). All those factors can affect nematodes directly. Certainly, unpredictable impacts of natural or artificial nature such as erosion (Urzelai et al., 2000), wildfire or windfall (Čerevková and Renčo 2009), heavy metal pollution (Valocká and Sabová, 1997a; Pen-Mouratov et al., 2008; Šalamún et al., 2011), agricultural practices (Rahman et al., 2007; Lišková and Renčo, 2007), acidification (Urzelai et al., 2000) may affect and may change (disturbance) the soil conditions and also the species richness, trophic structure, succession status and structure of nematode communities.

The overgrown plant communities can most likely affect the composition of natural nematode communities in grasslands and presence of particular nematode species. Plants can affect soil organisms through several mechanisms (Wardle, 2002). They affect soil biota (Neher, 1999) directly by generating inputs of organic matter above and belowground and indirectly by the physical effect of shading, soil protection and uptake of water and nutrients by roots. Energy and nutrients obtained by plants eventually become incorporated in detritus that provide the resource base of a complex of soil food web. For example, plant roots exude amino acids and sugars that serve as a food source for micro-organisms (Curl and Truelove, 1986). Each plant root can be a food source for plant parasitic nematodes, dead root system is a food source for bacteria and fungi (decomposition processes), bacteria and fungi are a food source for bacteriovorous and fungivorous nematodes, nematodes are a food source for predators etc. It is known and clear that plant parasitic nematodes are directly affected by the higher plants at a site (Yeates, 1999). For host-specific nematodes, such as *Heterodera* or *Globodera* the correlation between host and nematode population growth is tight.

According to Moore et al., (1988) three basic energy pathways exist in the soil: by plants, bacteria and fungi. To root pathway belong primarily to herbivores such as pathogenic fungi, bacteria, nematodes, protozoa and their consumers. Bacterial pathway is associated with saprophytic and pathogenic bacteria and their relevant consumers (protozoa, bacteriovorous nematodes). Fungal path involves saprophytic, mycorrhizal and pathogenic bacteria and their particular consumers (fungivorous nematodes, oribatid mites and springtails). The root, bacterial and fungal pathways unite at a higher level in the food chain, i.e. omnivores and predators. Many nematodes are omnivores and feed on a different types of food depending of environmental conditions and food availability (algae, fungi, bacteria, small rotifers enchytraeids and small nematodes). Predators relate to secondary, tertiary and quaternary consumers, including nematodes, beetles, spiders etc. On the other hand, the nematodes are an important food resource of many soil organisms (e.g. arthropods) (Walter, 1988).

The influence of plant types on nematode communities was demonstrably confirmed by Wasilewska (1995) when investiging the differences in development of soil nematode communites in single and multi species grass experimental treatments. She concluded that floristically diverse community supports a more mature stage of nematode succession, whereas monoculture is subjected to greater parasitization of plant parasitic nematodes. This result was confirmed by De Deyn et al. (2004). They stated that plant species diversity and plant identity affected the diversity of nematodes as well. Results showed that the nematode diversity varied more between the different plant species than between different levels of plant species diversity, so that plant identity is more important than plant diversity. Nevertheless the nematode diversity in plant mixtures was higher than in any of the plant monocultures, due to the reduced dominance of the most abundant nematode taxa in the mixed plant communities.

Similarly, Viketoft (2008) or Viketoft et al. (2009), observed that the identity of the plant species affected both the total abundance of nematodes and the nematode community composition. The number of bacterial-feeding nematodes was greatest under grasses and legumes and was positively correlated with shoot biomass and negatively with root biomass. The response of the plant feeding nematodes, which differed in abundance under both the investigated legume and the forb species, suggests that the identity of the plant species is more important than the plant functional group, which is probably related to differences in plant secondary metabolites. De Deyn et al. (2004) also detected that plant species identity marginally affected the abundance of the plant-associated and fungal feeding nematodes and did not affect the higher trophic level feeding groups (omnivores, predators). De Deyn et al. (2004) also found that ecto- and semi-endoparasites dominated in the root zone of the grass species, while in the forb species sedentary endo- and ectoparasites were most abundant, especially in *C. jacea* and *L. Vulgare*.

However, based on the data of a seven-year experimental study on grassland fields (Sohlenius et al., 2011) it was stated that nematode diversity was little affected by plant diversity, though several species such as the obligate plant feeders *Tylenchorhynchus dubius*, *T. maximus* and *Pratylenchus* spp. occurred in higher numbers in monocultures than in mixtures of several plant species. The abundance of some bacterial- and fungal-feeding nematode taxa seemed also to be directly influenced by the plant species identity. On the other hand, in natural (non-experimental) conditions, the strong plant effect on the soil nematode communities and their changes in particular site was observed in several studies. For example, Jackson et al. (2002) observed that the development of woody vegetation on grasslands alter nematode communities. Sites with recent woody plant invasions have reduced nematode species richness, particularly among root-parasite species. Biederman and Boutton (2009) found that biodiversity and trophic structure of soil nematode communities were altered by woody plant invasion of grassland as well.

In conclusion, as stated by De Deyn et al. (2004) resource quality is more important than resource quantity for the diversity of soil food web components and that plant species identity is more important than plant diversity per se.

NEMATODES IN NATURAL AND SEMI-NATURAL GRASSLANDS

The spectrum and denstities of soil nematode species in natural and semi-natural grassslands has been extensively studied in the 20th century worldwide and results approved that grassland, meadow and pasture soils contain a rich, abundant and diverse nematode fauna (Wasilewska, 1994; Popovici, 1998; Hodda and Wanless, 1994b; De Goede and Bongers, 1998). The nematode abundance in some natural European grasslands ranges between 0.4 x 10^6 and 12 x 10^6 individuals/m^2 (Gerber, 1985; Wasilewska, 1994; Ilieva, 1998; Popovici and Ciobanu, 2000) related to high index of generic and species diversity (Wasilewska, 1994; Popovici and Ciobanu, 2000).

A total of 74 nematode species, with bacteriovorous and plant parasitic nematodes being the dominant species, was recorded in meadows of Poland by Wasilewska (1976). In nematode communities the species *Tylenchorhynchus dubius, Paratylenchus curvitatus, Acrobeloides nanus* and *Plectus parvus* prevailed. In Romania, Suatu Botanical Reserve where steppe dominated, Ciobanu and Popovici (1999) found 97 nematode taxa from 74 genera. The dominant nematode taxa were *Rotylenchus, Paratylenchus, Tylenchorhynchus, Filenchus* and *Coslenchus.* High abundance and diversity characterized the nematode communities of 36 Romanian grasslands (Popovici and Ciobanu, 2000). Authors found 145 nematode species belonging to 121 nematode genera. The genera with maximum occurrence were *Filenchus*, *Rotylenchus*, *Paratylenchus*, *Acrobeloides* and *Plectus*. Dmowska (2000) identified 140 nematode species from 71 genera in the rhizosphere of subalpine meadows and evaluated the dominance of nematode taxa according to different soil types and climates of the investigated regions of Pyrenees. In Sweden semi-natural grassland, 79 nematode genera were found with the dominance of *Acrobeloides*, *Helicotylenchus*, *Prismatolaimus* and *Aphelenchoides* (Viketoft, 2007).

In the Slovak Republic, 46 nematode genera in meadows with the dominance of plant feeders *Helicotylenchus* were reported by Valocká et al. (2001); 66 nematode species from 49 genera in meadows with the dominance of plant parasite were recorded by Lišková and Čerevková (2006). Similarly, Zolda (2006) in the semi-natural steppe grasslands of Eastern Austria found 58 nematode genera where *Acrobeloides*, *Anaplectus*, *Heterocephalobus*, *Prismatolaimus*, *Aphelenchoides*, *Aphelenchus*, *Tylenchus* and *Pratylenchus* dominated, but in natural meadows of National Park Neusiedler in Austria Zolda (2002) found only 21 nematode genera with most abundant *Eumonhystera*, *Anaplectus* and *Acrobeloides* nematodes, all bacteriovores. Stamou et al. (2005) recorded only 39 nematode genera in Greek Mediterranean grassland, though in grasslands of the White Carpathians (Slovak-

Czech border) 155 nematode species from 86 genera were repoerted by Háněl and Čerevková (2006) with the dominance of bacteriovorous and plant parasitic nematodes.

As it was mentioned above, the nematode abundance, number of genera and species vary with site and time. Those differences are most likely related to various factors (ecological, geographical and environmental) as well as to differrent soil sampling and nematode extraction techniques used by researchers (Mc Sorley, 1987; Neher et al. 1995; Mc Sorley and Frederick, 2000). Advances in microscopic technology leads to a new classification of nematodes, a description of new genera and species, reclassifying already described species into new genera, families etc., which can increase the number of recorded species in similar ecosystems.

The trophic structure of soil nematode communities in European grasslands seems to be very similar, due to a dominance of bacteriovorous and plant parasitic nematodes. Certainly, this is derived from the dominant genera, which can therefore be regarded as common inhabitants of European grasslands such as *Acrobeloides*, *Plectus*, *Prismatolaimus*, *Tylenchus*, *Filenchus*, *Pratylenchus* etc. (Nagy, 1998; Valocká and Sabová, 1997b; Zolda, 2006). A dominance of plant parasitic nematodes is related to great plant diversity in natural grassland forming food base for plant-feeding nematodes however, they can by a serious threat to man-made grasslands as well.

NEMATODES IN AGRICULTURAL, MAN-MANAGED OR MONOCULTURE GRASSLANDS

The structure of soil nematode communities in the soil ecosystems of man-made grasslands is mainly influenced by human activities or management methods (grazing, fertilization, grass species grown etc.).

As stated by Bardgett et al. (1998), grazing of grassland by large herbivores can have substantial indirect effects on soil organisms, and might have significant impact on plant productivity and community structure. It was confirmed by Wang et al. (2006) who reported that cattle grazing increases microbial biomass and alters soil nematode communities in subtropical pastures. They found that microbial C and N were greater in grazed than in ungrazed plots. Effects of grazing varied among nematode genera. Most genera of colonizer bacterivores were decreased by grazing, but more persistent bacterivores such as *Euteratocephalus* and *Prismatolaimus* were increased, as were omnivores and predators. Abundance of omnivores, predators, and especially fungivores were strongly associated with C mineralization potential. The authors also suggested that grazing have impact on soil nematode communities but at a small local scale (1m x 1m), because in the previous study on the same pastures, relatively few effects of cattle density on nematode population levels were observed when relatively large (10m x 10 m) plots were used (McSorley and Frederick, 2000). Mulder et al. (2003) reported that most omnivorous and predatory genera are highly sensitive to livestock densities, and that heavy grazing reduced the abundance of most nematodes except for *Chiloplacus* and *Thonus*. Higher numbers of persisters indicated that grazing resulted in a more structured nematode community. Bardgett et al. (1997) recorded decrease in nematode abundance after cessation of grazing in upland pasture, which they refered to decreasion in microbial abundance as a food source for many microbivorous nematodes as well. Similar, Freckman et al. (1979) observed a greater abundance of nematodes (particular bacterivores) in a grazed than in ungrazed grassland. Nematode communities often reflect grazing treatments by shifts in their trophic group composition. In several studies, plant-feeding and fungal-feeding nematodes have been recorded at higher densities in grazed sites (Merrill et al., 1994; Freckman and Huang, 1998), unlike the studied of Wang et al. (2006) or Mulder et al. (2003) who suggested that heavy grazing was detrimental to herbivorous nematodes, especially for Criconematidae. Similarly to it, in the study of Zolda (2006) plant-feeding nematodes (*Pratylenchus*) and root-fungal feeding nematodes (*Tylenchus* and *Coslenchus*) showed higher abundance values at ungrazed sites.

However, Stanton et al. (1984) found no changes in nematode communities between long-term grazed and ungrazed treatments. Zolda (2006), having conducted the study on four grazed and four ungrazed semi-natural steppe grassland sites, reported that grazing had little or no effect on nematode community parameters, but total abundance was higher at ungrazed areas. This phenomenon (Zolda, 2006) was attributed to a lower herbage production and an altered root quantity and quality due to grazing, which has a direct effect on plant-feeding nematode populations. Significant differences were found in the percentage of omnivorous nematodes, the sum of the maturity index, number of genera and Simpson's index of diversity at a continuosly grazed pasture.

It seems that studies on the influence of domestic livestock on soil biological processes and communities in managed grasslands are contradictory and suggesting that more studies need to be conducted under a range of environmental and climatic conditions so that reasons for discrepancies among current studies may be resolved Wang et al. (2006).

The application of organic and inorganic fertilizers to grasslands has been shown to affect nematode populations as well. Such, application of organic manures to grassland increases nematode numbers, due to increased substrate availability, plant growth and altered soil conditions (Bardgett and Griffiths, 1997). Similarly to it, dairy manure slurry application increased microbial production supported substantially increased production of microbivorous protozoa and nematodes (mainly bacteriovorous) that are indicative of enhanced microbial turnover and flux of nutrients through the soil food web (Forge at al., 2005). But inorganic fertilisers have variable effects; one reduced (Sohlenius and Wasilewska, 1984) and second enhanced nematode populations (Beare et al., 1989). On the other hand cessation of fertiliser application has been shown as great factor of changes in plant-feeding nematode community (Verchoor et al., 2001).

Several studies focused on study of nematode community changes in man-disturbed grassland left to natural succession. Háněl (2003) investigated the spectrum and densities of soil nematode species in an extensively managed sub-thermophilous meadow and in conventionally managed and abandoned fallow fields left to natural succession. Extensively managed sub-thermophilous grassland had a great diversity of soil nematodes (115 species and 71 genera) and the dominant trophic groups were root-fungal feeders and bacterivores followed by fungivores and omnivores. Ploughing the meadow soil and 30 years of intensive agricultural practices have changed the nematode fauna. Only 66 species and 41 genera were found in the cultivated field and about one-half of all nematode individuals were bacterivores. During a two-year period when the arable field soil was abandoned and left to natural secondary succession, the abundance and the dominance of omnivores increased whereas the dominance of bacterivores decreased. The number of species increased to 73, the number of genera to 46. These results indicate that many nematode populations are able to recover from the cropping stress in a relatively short period of several years of natural secondary succession after cessation of the intensive agriculture (Háněl, 2003).

Wasilewska (1994) found that trophic structure of nematode communities and generic diversity was more dependent on habitat conditions other than meadows age. But Maturity index positively correlated with the age of the meadows, thus with succession from new-established ley to permanent meadows.

Successive changes of nematode communities inhabiting agricultural soil by sown grass mixtures to this soil (regional and commercial) were investigated by Háněl (2001). He reported that in the grass mixture plots total nematode abundance increased rapidly, whereas in the field it remained relatively low. Several microbivores increased rather quickly (*Panagrolaimus*, *Paraphelenchus*, *Aphelenchus*, *Aphelenchoides*) while an increase in plant parasite populations was slower. The development of nematode communities suggests a great role of accumulation of above-ground and below-ground litter supporting microflora and microbial feeding fauna in an initial stage of meadow restoration. In older stages, an increase of root density is expected with multiplication of plant parasitic nematodes (Háněl, 2001), which are important in terms of the green mass production. They include in particular endoparasitic nematode of genus *Pratylenchus* (Ustinov and Zinovev, 1963), ectoparasitic nematodes of the genus *Helicotylenchus* (Chastagner and McElroy, 1984), *Criconema* and *Criconemoides* (Evans et al. 1993; Lišková et al., 2004), *Paratylenchus* (Brzeski, 1995), *Longidorus* and *Xiphinema* (Evans et al. 1993; Lišková and Brown, 2003), cyst forming nematodes of the genus *Heterodera* (Sturhan and Lišková, 2004). Wasilewska (1995) recorded *Pratylenchus* as dominant genus in grass monoculture as well.

Contrary to it, Li et al. (2007) investigated successive changes of nematode communities during the transformation of grass and shrubs field to tea plantation. Nematode abundance, generic richness, diversity, and proportions of bacterivores and fungivores were higher at each stage of GT transect (grassland to tea plantation) than were those at the equivalent stage of ST transect (shrubland to tea plantation), whereas lower values of maturity index and nematode channel ratio occurred at each stage of GT transect relative to ST transect. This suggested that the original vegetation types may affect the following nematode community development; however, the effects can decrease over time

Soil Nematodes and their Use as Indicator of Soil Ecosystem Status

Developments in the ecology of soil nematodes has progressed thorugh several stages during the last couple decades. The studies relating to nematode communities and their role in ecosystems have given information on nematodes in different types of natural or human controled ecosystems. But sufficient information about the soil nematode communities in the natural ecosystems or ecosystems that are relatively undisturbed by human activities can provide useful insights into soil conditions in terrestrial habitats (Hodda and Wanless, 1994a, b). Since Bongers (1990) has established Maturity index (MI) for ecological measuring of environmental disturbance by nematode species composition, the analysis of nematode community structure has been more frequently used as a tool in ecological studies.

Nematodes were designed for their use as indicators due to several behaviours that they have evolved. They are frequently the most numerous component of the microfauna and are considered to be the most abundant and diverse invertebrates present in the soil (Yeates, 1979). Nematodes belongs to the simplest metazoa, occur in any environment that provides a source of organic carbon, in every type of soil, under all climatic conditions and in habitats that vary from authentic to extremelly polluted (Bongers and Ferris, 1999). In soil, nematodes live in capillary water and their permeable cuticle provides direct contact with their microenvironment. They do not rapidly migrate from stressful conditions and many species survive dehydratation, freezing or oxygen stress. Nematodes occupy key positions in soil food webs. They feed on most soil organisms and are food for many others. They can also influence vegetation succession.

Because their feeding habits are clearly based on shape and structure of mouth (Fig. 1) their trophic roles are readily inferred (Bongers and Ferris, 1999), although the deeper investigations confirmed that the species belonging to the same genus can have a different feeding habits (Tylenchida). Wasilewska (1971) and Yeates *et al.* (1993) separated soil nematodes to trophic groups, which builds on the original ecological classification (Paramonov, 1952). The trophic categories and functional guilds have been used to develop a number of indices for assessing nematode diversity, being weighted in favour of the rarer nematode types, whereas other indices have a numerical weighting (Pattinson *et al.*, 2004). Representation of trophic groups in the community, ecological and diversity indices serve as indicators for assessing soil environment.

Bacterial feeders: feed on the products of decomposition of organic residues in soil. Indicate the increased soil microbial activity in particular for increasing the nitrogen content;

Fungal feeders: feed on fungi and are dependent on soil pH. Induce increased acidity of the environment and the predominance of putrefactive processes in the soil;

Plant feeders: obligatory plant parasitic nematodes are endo-ectoparasitic nematodes of plant roots;

Root-fungal feeders: facultative plant parasitic nematodes with nonspecific pathogenic effect;

Omnivores: free living nematodes; feed mainly on residues of soil organic matter;

Predators: feed on another soil organisms, including nematodes.

Nematodes respond rapidly to disturbance and enrichment: increasing microbial activity leads to changes in the proportion of bacterial feeders in a community (Bongers and Ferris, 1999). Though, nematodes represent a relatively small weight of biomass in the soil, their occurrence inside the several trophic groups is noticeably important in the soil environment (Barker and Koenning, 1998).

Each soil sample contains thousands or millions diverse nematode specimens. For instance, Zlotin (1969) found five million and Wasilewska (1976) two million nematode individuals in grassland soils. In addition to that, nematodes are easy to extract from the soil using simple extraction procedures (Ritz and Trudgill, 1999).

As nematodes are transparent, their diagnostic internal features can be seen without dissection (Bongers and Ferris, 1999).

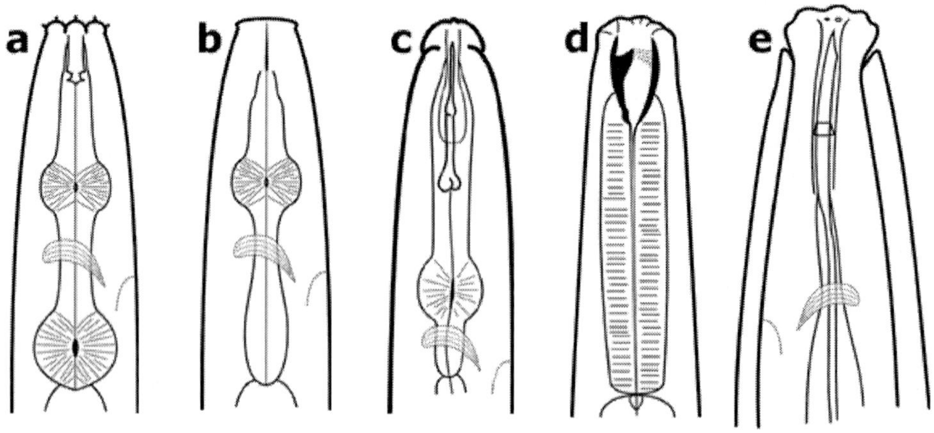

Figure 1 Classification of soil nematodes into different feeding groups based on their mouth structures. (a) bacterial feeder, (b) fungal feeder, (c) plant feeder, (d) predator, (e) omnivore. (Zaborski, 2009)

Their main drawback is the problem of identification to the species level, but this has been circumvented by the use of trophic group categories and functional guilds which require reduced taxonomic skills (Pattinson *et al.,* 2004). In conclusion, nematodes pose an excellent biological material because they are abundant, ubiquitous, easily collected and extracted from soil and their diversity is astonishingly high.

Ecological Indices

Since the 1980's, there has been a rising interest to use soil nematode communities as indicators for environmental monitoring of terrestrial ecosystems (Freckman, 1988; Bongers, 1990). At first, the Maturity Index (MI) for free-living nematodes and Plant Parasitic Index (PPI) for plant parasitic nematodes has been developed by Bongers (1990), at second the modifications of MI have been proposed as MI2-5 (Bongers and Korthals, 1993); ∑MI (Yeates, 1994) and ∑MI2-5 (Neher and Cambell, 1996). Consequently, the proportions of nematode trophic groups and ecological indices were established (Bongers and Korthals, 1995, Wasilewska, 1997) or diversity indices were employed (Shannon-Weaver 1949; Heip *et al.*, 1988; Yeates *et al.*, 1993). However, the diversity indices, generic or trophic group abundance has been used for analyses of anthropogenic effects on nematode communities structure and function before any establishment of ecological indices (Yeates, 1982; Sohlenius and Wasilewska, 1984; Hendrix *et al.* 1986; Juma and Mishra, 1988)

Calculation of maturity indices (MI, PPI, ∑MI, MI2-5, ∑MI2-5) is based on the fact that nematode species are divided according the length of the lice cycle, reproduction ratio, tolerance to changes in soil conditions into a five *c-p* (colonizers-persisters) groups (1-5). *C-p*=1 represents r-strategists or colonizers, *c-p*=5 represents K- strategists or persisters. Each nematode taxon, usually at family level, is classified into one of the five *c-p* groups. Genera and species within a taxon have the same *c-p* value as their family.

For the terrestrial and freshwater nematode taxa, the following groups can be distinguished (Ferris and Bongers, 2009):

***c-p* 1** - nematodes with a short race time and a large proportion of the body occupied by gonads which produce many small eggs. Population growth under food-enriched conditions is exceeding. The nematodes are primarily bacterial-feeders with high metabolical activity. They are tolerant of pollutants and of products of organic matter decomposition. These enrichment opportunists form dauer larvae when microbial biomass and activity decreases.

***c-p* 2** - nematodes with a short race time and relatively high reproduction rates, although lower than are those in *c-p*1 and they are slower to respond to environmental enrichment than *c-p*1 nematodes. These nematodes do not form dauer larvae and occur in all environments, including those in which resources are abundant and those in which resources are scarce. They are very tolerant of pollutants and other disturbances and include bacterial and fungal feeders and some predators.

***c-p* 3** - nematodes with longer generation time than *c-p*2 nematodes and greater sensitivity to disturbances. They include bacterial and fungal feeders and some predators.

***c-p* 4** - small dorylaims and the large non-dorylimids, characterized by a long generation time, permeable cuticle and high sensitivity to pollutants. The non-predators in this group are relatively sessile whereas the predators actively seek prey. They include larger predators, smaller omnivores and some bacterial feeders.

***c-p* 5** - large dorylaimid nematodes with long life spam, low reproduction rates, low metabolic activity and slow movement, producing a small number of large eggs. They have permeable cuticle and are very sensitive to pollutants and other disturbances. This group is composed of the larger omnivores and predators.

The maturity index (MI) is estimated without parasitic nematodes and can show a rating of environemtal disturbance, with smaller values being indicative of a more disturbed and enriched environment and larger values characteristic of a less disturbed and stable environment. Bongers (1990) also proposed two kinds of changes in the composition of nematode community after disturbance. A type I response is found when perturbations result in an increased number of taxa with low *c-p* values and others do not, for example effect of eutrophication, fertilization and some physical stresses, because of those perturbations increase in nutrient availaibility. A II type of response (less often documented) is found when disturbance results in an absolute decrease among most taxa. It seems that this type of response can be anticipated at long-term effect of disturbances where toxic stress is attended with condition of low food availaibility (Korthals *et al.,* 1996). Moreover, environmental disturbances can be classified in many ways (Neher *et al.,* 2005). Firstly, disturbances may be classified by type, i.e., chemical, physical, or biological, which alter nematode communities in qualitatively and quantitatively different ways. Secondly, disturbance can be described by its characteristics, e.g., intensity, frequency, regularity and magnitude (Dyer and Letourneau, 2003). Thirdly, disturbances may be seasonal or otherwise cyclical. Fourthly, disturbances may be large or small in the impact scale, e.g., tree-fall or clear-cut. Overall, a disturbance regime may be specific to an ecosystem, geographic location or local climate, e.g., nutrient and pest management in annual agricultural systems, grazing pressure (e.g., species, stocking rate) on pastures, and harvest method of forests. Apparently, different functional groups and genera are more or less tolerant to different 'modes' of disturbance. Although maturity indices generally have the ability to respond to specific disturbances, responses have not been fully consistent across regions, ecosystems, seasons or other conditions (Neher *et al.*, 2005)

The MI decreases with increasing microbial activity as well. For example, the values of MI recorded in hop fields (pernamently cultivated and fertilized by cow manure) ranged from 1.5 to 2.6 (Lišková and Renčo, 2007) and can be considered as a more disturbed ecosystem than e.g. permanent meadows and pastures with MI 3.4, resp. 3.5 (Čerevková, 2006) or MI 3.6 (Háněl and Čerevková, 2006) (Table 1).

Plant Parasitic Index (PPI) is comparable to the MI but calculated only for the plant parasitic nematodes because their abundance is determined by the potency of their host plants and positively correlated with primary production (Bongers, 1990). Higher values would thus indicate nutrient-enrichment where nematodes with a long life cycle and low fecundity prevail but those ecosystems are sensitive to environmental change. Low values, on the contrary, indicate a les mature ecosystem where nematodes with short life cycle and high reproductive ratio prevail but those ecosystems are tolerant to environmental disturbance (Wasilewska, 1997, Neher, 2001). In a study of Šalamún *et al.* (2011) at heavy metal polluted grasslands, the PPI values decreased with the decreasing of heavy metal contents in soil. Although the decrease was not significant, it was probably due to two possible different ways of stress. The first stress could be represented by the heavy metals pollution of the soil and its negative impact on vegetation (Yeates *et al.,* 1994) and the second could be relatively high levels of nitrogen in the soil, which has negative influence on plant feeders and enhanced their suppression (Renčo *et al.*, 2009). The vlaues of PPI may (Neher and Campbell, 1994) or may not (Bongers *et al.,* 1997) positively correlate with MI. However, the data of conducted studies, regardless of the ecosystem type, suggest that an increase in the populations of Tylenchidae with *c-p* 2 decreased the values of PPI as it was also stated by Háněl and Čerevková (2006). That is in agreement with with the several other studies, e.g by Háněl (1996) where the *Filenchus* dominated, Wasilewska (1994) where the *Tylenchus* and *Filenchus* dominated, all related to low PPI (about 2.0- 2.2). Contrary to it, the values of PPI > 3.0 where the *Pratylenchus* (*c-p* 3) dominated or values of PPI = 3.0 where plant parasitic nematodes of *c-p* 3 dominated were observed by Wasilewska (1995), Brmež *et al.* (2006) or Lišková and Renčo (2007) (Table 1).

Bongers and Korthals (1993) were the first authors to discuss the modification of MI and they have established MI2-5. The index is identical to MI but excludes the *c-p*1 enrichment opportunists. The index was derived during studies of the relationship between MI and cooper concentrations under agricultural soils. It was evident that there was a strong relationship between decrease in higher *c-p* value nematodes and pollution stress while the *c-p*-1 nematodes responded to the presence of decomposing organic matter. Korthals *et al.* (1996) stated that the MI2-5 values give a much better response to disturbance than MI, especially in heavy metals polluted soils. However, in the study of Šalamún *et al.* (2011) at natural grassland, MI2-5 index continuously decreased with increasing distance from pollution source related to decreasing of heavy metal contents. The authors have ascribed the decreasing trend to different proportion of opportunistic nematodes and persisters at sites and negative correlation of MI ($P<0.01$) with the content of C_{ox} as a source of organic matter and increasing concentration of Cr. These two facts together cause probably the lowest MI2-5 value at the farthest site from a metallurgical plant. Similar response under enriched condition was experimentally proved by Nagy (1999) and Nagy *et al.* (2004).

Later, Yeates (1994) proposed the modification of the MI and established the ∑MI which is equivalent to the Total MI (Wasilewska, 1994). ∑MI is calculated using all taxa in a nematode community and the *c-p* scale of 1-5 (Bongers, 1990) is used. This serves the same purpose as MI and reacts similarly in natural and semi-natural conditions and Bongers (personal communication) assumed that in nutrient-rich agroecosystems a well represented group of plant parasitic nematodes may mask the changes in value of the index.

In conclusion, the MI and PPI are two indices which respond in opposite directions at an increasing nutritional status of the soil (Bongers *et al.,* 1997). Authors assumed that the MI increases as a result of an increasing proportion of enrichment opportunists, the PPI increases with an increased carrying capacity for plant parasitic nematodes. Maturity indices can be used as a measure of the ecological succession status of a soil nematode community

The PPI/MI ratio was established as sensitive parameter in the monitoring of agro-ecosystems. Bongers and Korthals (1995) stated that the MI decreases and the PPI increases with the increasing nutrient status. However, researchers used this ratio for evaluation of soil status of natural grasslands (Ciobanu and Popovici, 1999; Čerevková, 2006; Zolda, 2006); natural but heavy metal polluted grasslands (Šalamún *et al.,* 2011) or natural forests as well (Čerevková and Renčo, 2009). As the authors of this ratio have stated (Bongers and Korthals, 1995) the PPI/MI value does not exceed 0.9 in the natural habitats, undisturbed conditions with the rich plant diversity and nutrient supply. Effects of slight nutrient disturbances are indicated by a value of up to 1.2. The value over 1.6 indicates an ecosystem with the outstanding balance of nutrient related to decrease in soil fertility. These arguments has been approved by researchers, because of the PPI values recorded in agricultural (intensively cultivated) mostly beyond the 1.0 (Ivezic *et al.,* 2000; Renčo, 2004; Brmež *et al.,* 2006; Lišková and Renčo, 2007) compared to those in natural grassland (Ciobanu and Popovici, 1999; Valocká *et al.,* 2001; Háněl, Čerevková, 2006) (Table 1).

The bacteriovorous to fungivorous nematode ratio (B/F) is known to be an important indicator of the decomposition pathway of organic matter in the soil food webs (Yeates, 2003). Values of this ratio suggest that organic matter was turnover by decomposers (fungi, bacteria), which proportionally correspond with the abundance of the above trophic groups (Wardle and Lavelle, 1997). High values of the ratio indicate higher contribution of bacteria to the decomposition and relatively quick breakdown of available organic matter in the soil (Wasilewska, 1997). Decreasing trend of the ratio indicates relatively slower rate of organic matter turnover due to fungal pathway of decomposition (Porazinska et al., 1999).

Similar information can be provided by the Nematode Cannel Ratio (NCR) which is constrained to have values between 1 (totally bacterial-mediated) and 0 (totally fungal-mediated) (Yeates, 2003) (Table 1).

Ferris et al. (2001) suggested that soil nematode communities be used to assess soil food-web by Enrichment and Structure Indices (EI, SI). EI is based on the expected responsiveness of the opportunistic guilds (bacteriovorus nematodes with c-p value equals one) to organic resources enrichment.

Table 1 The values of ecological indices (their proportion) and B/F ratio recorded in different ecosystems

Ecosystem	Indices, ratios						References
	MI	PPI	PPI/MI	∑MI	MI2-5	B/F	
Grass stands							
Semi-natural grassland	1.8	2.5	-	-	3.5	-	Urzelai et al. 2000
Semi-natural grassland	2.4-2.5	2.5			2.5-2.6	5.3-5.7	Viketoft, 2007
Meadows	1.9-2.2	2.1-2.8	-	-	-	-	Wasilewska, 1997
Meadow	3.0-3.1	2.2	-	-	-	-	Háněl, 1995
Meadows (polluted)	2.4-2.8	2.4-2.6	0.9-1.0	-	2.6-3.2	1.8-6.2	Šalamún et al. 2011
Meadows (natural)	2.5-3.6	2.4-3.0	0.8-1.0	2.6-3.1	-	-	Háněl, Čerevková, 2006
Meadows (natural)	2.1-3.7	3.0-3.9	0.8-1.5	-	-	0.6-14.2	Lišková, Čerevková, 2005
New meadows	2.5	2.5	0.9	-	-	6.3	Čerevková, 2006
Ley meadows	1.6-2.6	2.3-3.0	-	-	-	-	Valocká, Sabová, 1997b
Permanent meadows	3.4	2.6	0.7	-	-	3.2	Čerevková, 2006
Grassland	2.7-3.4	2.5-2.6	0.8-1.0	-	-	-	Ciobanu, Popovici, 1999
Grassland	2.0-2.9	2.1-2.9	-	-	-	0.2-3.7	Stamou et al. 2005
Grassland	2.9-3.1	2.7-2.8	0.9-1.1	-	-	-	Valocká et al. 2001
Grass monoculture	2.9	3.1	-	-	-	-	Wasilewska, 1995
Grass mixed culture	3.0	3.0	-	-	-	-	Wasilewska, 1995
Grassland –steppe	2.7-3.4	2.5-2.6	0.8-1.0	-	-	-	Ciobanu, Popovici, 1999
Semi-natural steppe	2.3-2.6	2.4-2.8	-	2.2-2.6	-	-	Zolda, 2006
Grazed grassland	2.3-2.4	-	-	-	-	-	Wang et al. 2006
Ungrazed grassland	2.2-2.5	-	-	-	-	-	Wang et al, 2006
Permanent pastures	3.5	2.7	0.8	-	-	2.1	Čerevková, 2006
Wetland	2.2	2.9	-	2.6	2.5	-	Neher et al. 2005

Ecosystem	Indices, ratios						References
	MI	PPI	PPI/MI	∑MI	MI2-5	B/F	
Forest							
Spruce forest	3.0	2.4	0.8	-	-	4.0	Čerevková, Renčo, 2009
Mixed forest	2.4	3.0	-	2.5	2.5	-	Neher et al. 2005
Mixed forest	2.6-2.8	-	-	-	-	-	Forge, Simard, 2001
Oak forest	2.2-2.7	2.0	-	-	-	-	Háněl, 1995
Acacia nilotica forest	2.7-3.7	0-3.55	-	-	-	-	Tomar, Ahmad, 2009
Agricultural soils							
Sugar beet	2.3	2.4	1.0	-	-	-	Renčo, 2003
Lucerne	2.6	2.6	1.0				Renčo, 2003
Hop monoculture	1.5-2.6	2.8-3.0	1.2-2.0	-	-	1.0-10.3	Lišková, Renčo, 2007
Cereals	2.4-2.5	2.5.-2.6	1.0.-1.1	-	-	-	Renčo, 2004
Soybean	2.7-3.3	2.5-2.9	-	2.7-3.2	2.7-3.2	-	Gomez et al. 2003
Wheat-potato	2.0-2.3	2.4	-	-	-	-	Háněl, 1995
Wheat	1.9-2.3	2.3-3.0	1.3-1.8	-	-	-	Brmež et al. 2006

Therefore, EI describes whether the soil environment is nutrient enriched (high EI) or depleted (low EI). SI represents an aggregation of functional guilds with c-p values ranging from 3-5 and describes whether the soil ecosystem is structured with greater trophic links (high SI) or degraded (low SI) with fewer trophic links. On the basis of trophic habits and life cycle authors assign individual families coefficients for the calculation of these indices. EI index includes families Rhabditidae, Panagrolaimidae, Diplogastetidae (Ba1), Aphelenchidae, Aphelenchoididae, Anguinidae (Fu2), and Cephalobidae (Ba2). Generally, the low values of EI reflect low abundance of Rhabditidae and dominance of Cephalobidae. SI index includes predators of group of Aphelenchidae (Ca2), and several other families such Tripylidae (Ca3), Diphtherophoridae (Fu3), Prismatolaimidae (Ba3), Dorylaimidae (Om4), Mononchidae (Ca4), Leptonchidae (Fu4), Thornenematidae and Quadsianematidae (Om5) and Discolaimidae (Ca5).

Graphic representation of the faunal profile on Enrichment Trajectory and the Structure Trajectory (Fig. 2) indicates status of soil community in:

Quadrant A – enrichment damaged, nitrogen-rich, decomposition of organic matter through bacterivores, low ration C/N, food-web disrupted.

Quadrant B – enrichment less damaged, nitrogen-rich, decomposition of organic matter through bacterivores and fungivores, low ration C/N, food-web formed.

Quadrant C – enrichment natural (un-damaged), balanced supply of nutrients, decomposition of organic matter by fungivorous, C/N proportion high, food-web structural.

Quadrant D- stressed enrichment, nutrient poor with prevalence of fungivores, C/N ratio high, food-web indicates a degradation of soil environment.

Further, Ferris et al. (2001) also proposed the Channel index (CI), which is a percentage of fungivores among the total fungivores and c-p one opportunist´s bacterivores. CI indicates predominant decomposition channels in the soil food web, a high CI (> 50 %) indicates higher proportion of fungal decomposition (fungal decomposition channels) and reflect the high relative abundance of c-p2 fungal feeders (Aphelenchoididae and Aphelenchidae) and the corresponding low abundance of c-p1 bacterial feeders (Rhabditidae and Panagrolaimidae) whereas low CI (< 50 %) suggests bacterial decomposition channels. Wasilewska (1997) suggested that high values of CI indicate a slow cycling of elements in the ecosystem because the breakdown of dead tissue by bacteria and fungi releases bio-elements at a rate slower than that caused by phytophages. This may also indicate that there is relatively greater mineralization of nutrients via the decomposition pathways than via consumption of primary production by the nematodes.

Nevertheless, an assessment of soil food-web in ecological studies using EI, SI and CI indices is rare (Table 2) since better results were noticeable only in longer term studies. It was confirmed by Neher *et al.* (2005) in a long-term study where values of EI index were greater in disturbed compared to undisturbed soils. Similarly, Liang *et al.* (2007) published results of a long-term study, where values SI were significantly higher in undegraded and improved grassland than in the degraded grasslands sites. Wang *et al.* (2006) found lower EI and higher MI and SI values in grazed plots than in ungrazed plots and suggested that a nematode community in the grazed plots composed more abundant nematodes with higher c–p values, thus a more structured community than ungrazed plots. Biederman and Boutton (2009) pointed out that values of SI suggest that ecological conditions relevant to nematode community structure have deteriorated following woody invasion of grassland, while De Deyn *et al.* (2004) recorded that the nematode functional diversity indices (SI, EI and CI) were not affected by plant species diversity. Likewise, plant species identity did not affect most nematode functional indices. Berklemans *et al.* (2003) in a 12-year study in the sustainable agriculture farming systems where conventional (CONV), low-imput (LOW) and organic (ORG) management treatments were compared found the differences among the treatments. EI and SI were generally lower and BI and CI higher in CONV than in LOW and ORG treatments. They found a significant crop effect on the community indices throughout the years.

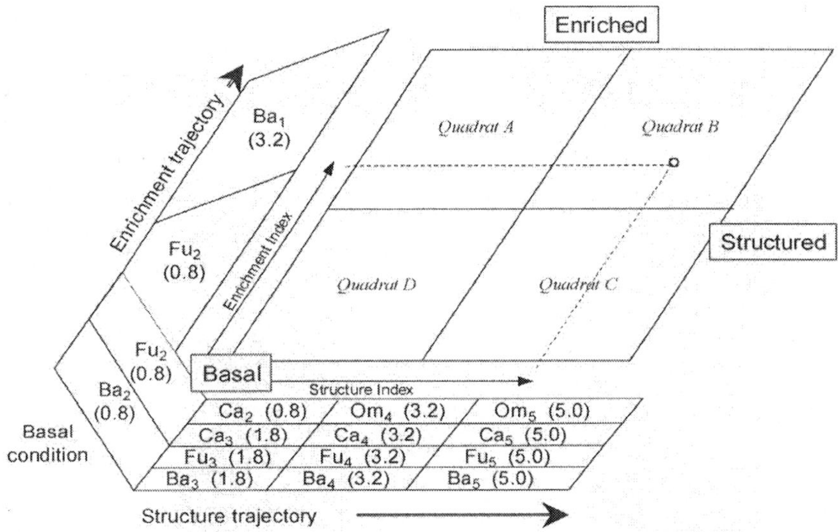

Figure 2. Functional guilds of soil nematodes characterized by feeding habit (trophic group) and by life history characteristics, according tor Bongers and Bongers, 1998. Indicator guilds of soil food web condition (basal, structured, enriched) are designated and weightings of the guilds along the structure and enrichment trajectories are provided, for determination of the Enrichment Index and Structure Index of the food web (Ferris *et al.* 2001).

Contrary to it, in a 12-month- study (Čerevková and Renčo, 2009) in spruce forest destroyed by windfall and wildfire the authors tried to assess changes in soil environment by EI and SI, but results were dubious, probably due to a short duration of the study and the changes above the ground had not yet affected the soil environment. Bulk soil samples from the forest undestroyed by windfall, destroyed by windfall and destroyed by windfall and subsequently by wildfire fell into Quadrat C – characterized as undisturbed with a moderate or higher C:N ratio and structured food web condition.

Diversity Indices

The nematode community diversity may be calculated at three levels of resolution: a) diversity based on abundance of individuals within each genus), b) trophic diversity based on abundance of individuals within each trophic group (trophic diversity), and c) diversity of genera within each trophic group (trophic richness) (Neher et al., 2004). The Shannon index diversity, Shannon index of trophic diversity (Shannon-Weaver, 1949), Trophic diversity (Heip et al., 1988) or Index diversity (Simpson, 1949) are most commonly used in combination with many others parameters to assess the nematode fauna in soil ecosystems and changes in nematode fauna resulting from ecosystem processes (Wasilewska, 1997) (Table 2).

Shannon-Weaver (H'gen, H´spp) and Simpson (λ) indices of diversity are mathematical expressions of taxonomic richness as well as the evenness of distribution of abundances among the taxa (Washington, 1984). Consequently, changes in index values can reflect changes in species richness, evenness, or both. Shannon-Weawer index is sensitive to rare species, while, Simpson index diversity weight common species. It was confirmed by Šalamún et al. (2011) in heavy metal polluted natural grasslands, where according to H´ index, site with the highest diversity was site the closest to the pollution source and with the lowest diversity was the site the farthest from the pollution source. According to λ, the value for diversity has the exact opposite trend, with the highest value at site the farthest and with the lowest value for diversity at site the closest to factory.

According to Wasilewska (1995) the H´index as well as taxa richness is much higher in habitats with the more diverse plant community. She found that values of H´index were significantly higher (p=0.01) in the soil with mixture of six grass species in comparison to grass monoculture, as was confirmed by several others studies. Lišková and Čerevková (2006) recorded the higher values of H´gen and H´spp indices in river banks compared to those in river adjacent meadows. Similarly, Čerevková and Renčo (2009) found that value of H´gen was higher in natural Lariceto-Piceetum forest compared to soil of destroyed Lariceto-Piceetum forest by windfall where grass *Calamagrostis villosa* being dominated. Popovici and Ciobanu (2000) investigated 36 grasslands in Romania, distributed at altitudes from 350 to 2270m a.s.l. and represented by 15 different plant associations developed on different soil types. They found that values of the H´index for nematode communities developed in rendzina, brown earth and acid brown soils had larger ranges (2.38-3.47) than those developed in lithosol, podzol and alpine meadows soils (2.65-3.13). However, clear differences between the different nematode communities of grassland of different altitudes and plant composition were not found. Neher and Campbell (1994) recorded that trophic diversity (Shannon) correlated positively with population levels of fungal-feeding, omnivorous and predacious nematodes, but negatively with population levels of bacterial-feeding nematodes.

Li et al. (2007) investigated successive changes of nematode communities during the transformation of grass and shrub field to tea plantation. They found that grassland had more nematode genera and a higher Shannon–Weaver diversity index value than shrubland because of great plant diversity in grassland. Similarly, the Trophic diversity index (TD) (Heip et al., 1988), which describes functional groups within the nematode communities, exhibited different values at the different investigation stations along the climatological gradient, was higher in GT transect (grassland to tea plantation) than in ST transect (shrubland to tea plantation) (Li et al., 2007).

On the other hand, the investigation of impact of heavy metals on nematode community structure (Li et al., 2006) revealed that H´index, TD index together with evenness index (J´) and index of dominance (λ) were sensitive to changes of soil pH and C/N ratio.

Table 2. The values of diversity indices, enrichment, structure and channel indices recorded in different ecosystems

Ecosystem	Indices							References
	H´spp	H´gen	TD	E	EI	SI	CI	
Grass stands								
Meadow	1.4-3.1	1.3-3.2	-	-	-	-	-	Lišková, Čerevková, 2005
Meadow	3.5-3.6	3.1-3.2	-	-	-	-	-	Háněl, 1995
Meadows	1.3-3.0	1.3-2.9	1.8-5.7	-	22-90	69-96	2-70	Háněl, Čerevková, 2006
New meadows	2.5	-	-	-	-	-	-	Čerevková, 2006
Permanent meadows	2.5	-	-	-	-	-	-	Čerevková, 2006
Grasslands	-	-	1.9-3.0	-	-	-	-	Valocká et al, 2001
Grassland	2.0	-	2.3	-	-	-	-	Urzelai et al. 2000
Grassland	2.2-3.0	-	-	75-92	-	-	-	Stamou et al. 2005
Grass monoculture	-	4.2	-	-	-	-	-	Wasilewska, 1995
Grass mixed culture	-	4.4	-	-	-	-	-	Wasilewska, 1995
Permanent pastures	2.2	-	-	-	-	-	-	Čerevková, 2006
Grazed grassland	-	-	4.1-7.0	-	45-48	7-34	54-77	Wang et al., 2006
Ungrazed grassland	-	-	4.3-9.0	-	50-56	3-17	36-68	Wang et al. 2006
Wetland	-	-	3.2-3.3	-	-	-	-	Neher et al. 2005
Forest								
Oak forest	3.0-3.9	2.5-2.8	-	-	-	-	-	Háněl, 1995
Spruce forest	2.3	2.1	-	-	-	-	-	Čerevková, Renčo, 2009
Mixed forest	-	-	2.0-3.2	-	-	-	-	Neher et al. 2005
Mixed forest	-	2.5-2.7	-	-	-	-	-	Forge, Simard, 2001
Acacia nilotica forest	-	1.4-2.6	1.0-3.5	-	0-50	27-100	0-100	Tomar, Ahmad, 2009
Agricultural soils								
Cereals	-	-	2.3-2.7	-	-	-	-	Valocká et al. 2001
wheat-potato	2.8-3.0	2.7-2.8	-	-	-	-	-	Háněl, 1995
Soybean	-	0.8-1.0	1.7-2.6	-	-	-	-	Gomez et al. 2003

CONCLUSION

It can be concluded that information about the soil nematode communities in the natural ecosystems or ecosystems that are relatively undisturbed by human activities, together with the indices which are calculated from nematode fauna analyses provide an excellent and useful insights into their soil conditions. On the other hand, the analyses of nematode community structures are more frequently used as a tool in ecological studies investigating the effects of various disturbances of overgrown vegetations and soil environments by natural or artificial characters. Together with the analyses of several other soil animals, analyses of the soil micro-fauna (bacteria, fungi), analyses of the plant communities and more conventional soil physical and chemical tests can provide a deeper understanding of how environmental (e.g fire, erosion) and anthropogenic disturbances (e.g. land management, clear-cut) impact on soil health. Thought, most studies indicated that land use changes or environmental changes markedly affect soil nematode community; these effects often appear to be idiosyncratic and case specific. In the future, deeper studies of nematode species biology and nematode species food habits will be required, as the differences between the same ecosystems in different world studies are clear. However, naturally occurring species of nematodes, ecological, soil and climatic conditions of studied soil ecosystems around the world will probably still be the reason why the data on the number of nematodes species or genera, species or genera diversity, and related differences in the values of ecological and diversity indices will be unequal. But in general, nematodes can be considered as reliable environmental indicators due to their obvious strong relationships with land management and aboveground vegetation.

REFERENCES

Bardgett, R.D., Cook, R. (1998) Functional aspects of soil animal diversity in agricultural grasslands. *Applied Soil Ecology* 10: 263–276.

Bardgett, R.D., Griffiths, B. (1997) Ecology and biology of soil protozoa, nematodes and microarthropods. In: van Elsas, J.D., Wellington, E., Trevors, J.T. (Eds.), *Modern Soil Microbiology*. Marcell Dekker, New York, pp.129-163.

Bardgett, R.D., Leemans, D.K., Cook, R., Hobbs, P.J. (1997) Seasonality in the soil biota of grazed and ungrazed hill grasslands. *Soil Biology and Biochemistry* 29: 1285–1294.

Bardgett, R.D., Wardle, D.A., Yeates, G.W. (1998) Linking above-ground and below-ground interactions: how plant responses to foliar herbivory influence soil organisms. *Soil Biology and Biochemistry* 30: 1867–1878.

Barker, K.R., Koenning, S.R. (1998) Developing sustainable systems for nematode management. *Annual Review of Phytopathology* 36: 165-205.

Beare, M.H., Blair, J.M., Parmelee, R.W. (1989) Resource quality and trophic responses to simulated throughfall: effects on decomposition and nutrient flux in a no-tillage agroecosystem. *Soil Biology and Biochemistry* 21: 1027-1036.

Berkelmans, R., Ferris, H., Tenuda, M., van Bruggen, A.H.C. (2003) Effects of long-term crop management on nematode trophic levels other than plant feeders disappear after 1 year of disruptive soil management. *Applied Soil Ecology* 23: 223–235.

Biederman, L.A., Boutton, T.W. (2009) Biodiversity and trophic structure of soil nematode communities are altered following woody plant invasion of grassland. *Soil Biology & Biochemistry* 41: 1943–1950.

Bongers, T. (1990) The Maturity Index: an ecological measure of environmental disturbance based on nematode species composition. *Oecologia* 83: 14–19.

Bongers, T., Bongers, M. (1998) Functional diversity of nematodes. *Applied Soil Ecology* 10: 239-251.

Bongers, T., Ferris, H. (1999) Nematode community structure as a bioindicator in environmental monitoring. *Trends in Ecology and Evolution* 14: 224–228.

Bongers, T., Korthals, G. (1993) *The Maturity Index, an instrument to monitor changes in the nematode community structure*. Summaries of the 45th International Symposium on Crop Protection, May 4, Gent, Belgium, pp. 80.

Bongers, T., Korthals, G. (1995) The behavior of maturity index and plant parasite index under enriched conditions. *Nematologica* 41: 286.

Bongers, T., Van Der Meulen, H., Korthals, G. (1997) Inverse relationships between the nematode maturity index and plant parasite index under enriched nutrient conditions. *Applied Soil Ecology* 6: 195 – 199.

Brmež, M., Ivezić, M., Raspudić, E. (2006) Effect of mechanical disturbances on nematode communities in arable land. *Helminthologia* 43: 117-121.

Brzeski, M.W. (1995) Paratylenchidae: Morphology of some known species and description of *Gracilacus bilineate* sp. n. and *G. vera* sp. n. (Nematoda: Tylenchulidae) *Nematologica* 41: 535-565.

Chastagner, G.A. McElroy, F.D. (1984) Distribution of plant parasitic nematodes in putting green turfgrass hosts of some graminaceous cyst nematodes. *EPPO Bulletin* 12: 399-411.

Ciobanu, M., Popovici, I. (1999) *Diversity of soil nematode communities in the Suatu Nature Reserve*. Studia Universitatis Babes-Bolyai, Biologia, XLIV, 1-2, pp. 45-51.

Curl E.A., B. Truelove (1986) *The Rhizosphere.* Springer-Verlag, Berlin, 288 pp.

Čerevková, A. (2006) Nematode communities in three types of grassland in the Slovak Republic. *Helminthologia* 43: 171-176.

Čerevková, A., Renčo, M. (2009) Soil nematode community changes associated with windfall and wildfire in forest soil at the High Tatras National Park, Slovak Republic. *Helminthologia* 46: 123-130.

De Deyn, G.B., Raaijmakers, C.E., van Ruijven, J., Berendse, F., van der Putten, W.H. (2004) Plant species identity and diversity effects on different trophic levels of nematodes in the soil food web. *Oikos* 106: 576-586.

De Goede, R., Bongers, T., (1998) *Nematode Communities of Northern Temperate Grassland Ecosystems*. Focus, Gießen, Germany.

Dmowska, E. (2000) Nematode communities in subalpine meadows in Central Pyrenees. *Annales Zoologici* 50: 211 – 220.

Dyer, L.A., Letourneau, D. (2003) Top-down and bottom-up diversity cascades in detrital vs. living food webs. *Ecology Letters* 6: 60–68.

Evans, K., Trudgill, D.L., Webster, J.M. (1993) *Plant Parasitic Nematodes in Temperate Agriculture*. Wallingford, CAB International, 648 pp.

Ferris, H., Bongers, T. (2009) Indices for Analysis of Nematode Assembladges. In. Wilson, M.J., Kakuoli-Duarte, T. (Eds.) *Nematodes as Environmental Indicators*. CAB International, Wallingford, UK, p. 124-145.

Ferris, H., Bongers, T., De Goede, R.G.M. (2001) A framework for soil food web diagnostics: extension of the nematode faunal analysis concept. *Applied Soil Ecology* 18: 13-29

Forge, T.A., Bittman, S., Kowalenko, C.G. (2005) Responses of grassland soil nematodes and protozoa to multi-year and single-year applications of dairy manure slurry and fertilizer. *Soil Biology & Biochemistry* 37: 1751–1762.

Forge, T.A., Simard, S.W. (2001) Structure of nematode communities in forest soils of southern British Columbia: relationships to nitrogen mineralization and effects of clearcut harvesting and fertilization. *Biology and Fertility of Soils* 34:170–178.

Freckman, D.W. (1988) Bacterivorous nematodes and organic-matter decomposition. *Agriculture, Ecosystems and Environment* 24: 195-217.

Freckman, D.W., Baldwin, J.G. (1990) Nematoda. In: Dindal, D.I. (Ed.) *Soil biology guide*. New York' John Wiley and Sons, p. 155– 200.

Freckman, D.W., Duncan, D.A., Larson, J.R. (1979) Nematode density and biomass in an annual grassland ecosystem. *Journal of Range Management* 32: 418–422.

Freckman, D.W., Huang, S.P. (1998) Response of the soil nematode community in a shortgrass steppe to long-term and short-term grazing. *Applied Soil Ecology* 9: 39–44.

Galvánek, D., Viestová, E. (2006) *Lúky Slovenska – sprievodca rozumným hospodárením*. Daphne, Bratislava, pp. 16 (In Slovak).

Gerber, K. (1985) Faunistische und ökologische Studie über die Nematoden einiger subalpiner Böden bei Bagdastein (Zentralalpen, Österreich). In. Franz, H. (Ed) *Veröffentlichungen des Österreichischen MaB-Programms*, Band 9, Universitätsverlag Wagner, Innsbruck, p. 113-131.

Gomez, G.S., Huang, S.P., Cares, J.S. (2003) Nematode community, trophic structure and population fluctuation in soybean field. *Fitopatologia Brasileira* 28: 18-25.

Háněl, L (1995) Secondary successional stages of soil nematodes in cambisoils of South Bohemia. *Nematologica* 41: 197-218.

Háněl, L. (1996) Comparison of soil nematode communities in three spruce forests at the Bobín Mount, Czech Republic. *Biológia (Bratislava)* 51: 485- 493.

Háňel, L. (1999) *Nematoda*. Folia Fac. Sci. Nat. Univ. Masaryk. Brun., Biol., 101, 67-77.

Háňel, L. (2001) *Short-term effects of grass mixtures on nematodes inhabiting agricultural soil in the Biíle Karpaty Mts*., Czech Republic, s preliminary study. 6th European Workshop no Soil Zoology. České Budějovice, April 23-25, p. 51-57.

Háněl, L. (2003) Recovery of soil nematode populations from cropping stress by natural secondary succession to meadow land. *Applied Soil Ecology* 22: 255–270.

Háněl, L., Čerevková, A. (2006) Diversity of soil nematodes in meadows of the White Carpathians. *Helminthologia* 43: 109 – 116.

Heip, C., Herman P.M.J., Soetaert, K. (1988) Data processing, evaluation and analysis. In. Higgins, R.P., Thiel, H. (Eds.) *Introduction to the Study of Meiofauna*, Smithsonian Institute Press, Washington DC, pp. 197–231.

Hendrix, P.F., Parmelee, R.W., Crossley, D.A., Coleman, D.C., Odum, E.P., Groffman, P.M. (1986) Detritus food webs in conventional and no-tillage agroecosystems. *Bioscience* 36: 374-380.

Hodda, M., Wanless, F.R. (1994a) Nematodes from an English chalk grassland: population ecology. *Pedobiologia* 38: 530–545.

Hodda, M., Wanless, F.R. (1994b) Nematodes from an English chalk grassland: species distributions. *Nematologica* 40: 116– 132.

Ilieva, Z.I. (1998) Nematode fauna of a grassland in the biospherical reserve Parangalitsa in Bulgaria. In. De Goede, R.G.M., Bongers, T. (Eds.). *Nematode Communities of Northern Temperate Grassland Ecosystems*. Focus, Giessen, pp. 131-138.

Ivezić, M., Raspudić, E., Brmež, M. (2000) Structure of nematode communities in different agroecosystems in Croatia. *Helminthologia* 37: 165-169.

Jackson, R.B., Banner, J.L., Jabbagy, E.G., Pockman, W.T., Wall, D.H. (2002) Ecosystem carbon loss with woody plant invasion of grasslands. *Nature* 418: 623–626.

Juma, N.C., Mishra, C.C. (1988) Effect of an annual and perennial crop on trophic group dynamics of nematodes. *Canadian Journal of Soil Science* 68: 101-109.

Korthals, G.W., Alexiev, A.D., Lexmond, T.M., Kammenga, J.E., Bongers, T. (1996) Long-term effect of copper and pH on the nematode community in agroecosystem. *Environmental Toxicology and Chemistry* 15: 979-985.

Li, Q., Jiang, Q., Liang, W. (2006) Effect of heavy metals on soil nematode communities in the vicinity of a metallurgical factory. *Journal of Environmental Sciences* 18: 323-328.

Li, Y., Feng, J., Chen, J., Wu, J. (2007) Original vegetation type affects soil nematode communities. *Applied Soil Ecology* 35: 68–78.

Lišková, M., Brown, D.J.F. (2003) Longidoridae (Nematoda: Dorylaimida) in the Slovak Republic. *Helminthologia* 40: 165-172.

Lišková, M., Čerevková, A. (2005) Nematode communities of river banks and adjacent meadows in the Slovak Republic. *Helminthologia* 42: 223-232.

Lišková, M., Renčo, M. (2007) Communites of free living and plant parasitic nematodes in hop gardens in Slovakia. *Helminthologia* 44: 80-86.

Lišková, M., Vovlas, N., Sasanelli, N. (2004) Criconematidae (Nematoda) in the Slovak Republic. *Helminthologia* 41: 161-170.

Liang, W.J., Zhong, S., Hua, J.F., Cao, C.Y., Jiang, Y. (2007) Nematode Faunal Response to Grassland Degradation in Horqin Sandy Land. *Pedosphere* 17: 611–618.

Mc Sorley, R. (1987) Extraction of nematodes and sampling methods. In. Brown, R.H., Kerry, B.R. (Eds.) *Principles and Practice of Nematode Control in Crops*. Academic Press, Orlando, pp. 13-47.

Mc Sorley R., Frederick, J.J. (2000). Short-term effects of cattle grazing on nematode communities in Florida pastures. *Nematropica*. 30, 211-221.

Merrill, E.H., Stanton, N.L., Hak, J.H. (1994) Responses of bluebunch wheatgrass, Idaho fescue, and nematodes to ungulate grazing in Yellowstone National Park. *Oikos* 69: 231–240.

Moore, J.C., Walter, D.E., Hunt, H.W. (1988) Arthropod regulation of micro- and mesobiota in below- ground detrital food webs. *Agricultural Ecosystems and Environment* 33: 419-439.

Mulder, C.H., De Zwart, D., van Wijnen, H.J., Schouten, A.J., Breure, A.M. (2003) Observational and simulated evidence of ecological shifts within the soil nematode community of agroecosystems under conventional and organic farming. *Functional Ecology* 17: 516–525.

Nagy, P. (1998) The nematode fauna of a dry continental steppe grassland at Lake Donge´r, Hungary. In: De Goede, R., Bongers, T. (Eds.), *Nematode Communities of Northern Temperate Grassland Ecosystems.* Focus, Gießen, Germany, pp. 181–186.

Nagy, P. (1999) Effect of an artificial metal pollution on nematode assemblage of a calcareous loamy chernozem soil. *Plant and Soil* 212: 35–43.

Nagy, P., Bakónyi, G., Bongers, T., Kádár, I., Fábián, M., Kiss, I. (2004) Effects of microelements on soil nematode assemblages seven years after contaminating an agricultural field. *Science of the Total Environment* 320: 131–143.

Neher, D.A. (1995) Measures of nematode community structure and sources of variability among and within agricultural practices. In. Collins H.P., Robertson, G.P., Klug, M.J. (Eds.) *The significance and regulation of soil biodiversity.* Kluwer Academic Publisher, pp. 187-201.

Neher, D.A. (1999) Soil community composition and ecosystem processes. *Agroforestry Systems* 45: 159-185.

Neher, D.A. (2001) Role of Nematodes in Soil Health and Their Use as Indicators. *Journal of Nematology* 33:161–168.

Neher, D.A., Bongers, T., Ferris, H. (2004) *Computation of nematode community indices.* Society of Nematologists Workshop, Estes Park, Colorado, 33 pp.

Neher, D. A. Campbell, C.L. (1994). Nematode communities and microbial biomass in soils with annual and perennial crops. *Applied Soil Ecology* 1:17-28.

Neher, D.A., Campbell, C.L. (1996) Sampling for regional monitoring of nematode communities in agricultural soils. *Journal of Nematology* 28: 196-208.

Neher, D.A., Wu, J., Barbercheck, M.E., Anas, O. (2005) Ecosystem type affects interpretation of soil nematode community measures. *Applied Soil Ecology* 30: 47–64.

Paramonov, A.A. (1952) Opyt ekologičeskoj klassifikcai fitonematod. *Trudy Gel'mintologičeskoj Laboratory* 6: 338-364.

Pattison, T., Badcock, K., Armour, J., Moody, P., Velupillai, R., Cobon, J., Lindsay, S., Gulino, L., Smith, L. (2004) Using nematodes as bioindicators for soil health in bananas. Proceedings of the International Soil Science Conference, 5–9 December, Sydney, Australia. Published on CDROM. Website *www.regional.org.au/au/soil.*

Pen-Mouratov, S., Shururov, N., Steinberger, Y. (2008) Influence of industrial heavy metal pollution on soil free-living nematode population. *Environmental Pollution* 152: 172-183.

Platt, H.M., (1994) Foreward. In: *The Phylogenetic Systematics of Free-living Nematodes*, S. Lorenzen, (Ed.) The Ray Society, London, pp 383.

Popovici, I. (1998) Structure of nematode communities in mountain grasslands from Romania. In: De Goede, R.G.M., Bongers, T. (Eds.), *Nematode Communities of Northern Temperate Grassland Ecosystems*. Focus, Giessen, pp. 221–240.

Popovici, I., Ciobanu, M. (2000) Diversity and distribution of nematode communities in grasslands from Romania in relation to vegetation and soil characteristics. *Applied Soil Ecology* 14: 27-36.

Porazinska, D.L., Duncan, L.W., McSorley, R., Graham, J.H. (1999) Nematode communities as indicators of status and processes of a soil ecosystem influenced by agricultural management practices. *Applied Soil Ecology* 13: 69-86.

Rahman, L., Chan, K.Y., Heenan, P.D. (2007) Impact of tillage, stubble management and crop rotation on nematode populations in a long-term field experiment. *Soil & Tillage Research* 95: 110–119.

Renčo, M. (2003) The Communities of nematodes in agroecosystems of sugar beet, cereals and lucerne in Eastern Slovakia. *Helminthologia* 40: 55-58.

Renčo, M. (2004) Communities of nematodes in cereals fields following sugar beet. *Helminthologia* 41: 109-112.

Renčo, M., Sasanelli, N., Šalamún, P. (2009) The effect of two compost soil amendments, based on municipal green and penicillin production wastes, on plant parasitic nematodes. *Helmithologia* 46: 190-197.

Ritz, K., Trudgill, D.L. (1999) Utility of nematode community analysis as an integrated measure of the functional state of soils: perspectives and challenges. *Plant and Soil* 212: 1-11.

Shannon, C.E., Weaver, W. (1949) *The mathematical theory of communication.* University of Illinois, Urbana, Illinois, USA.

Simpson, E. H. (1949) Measurement of diversity. *Nature* 163: 688 p.

Sohlenius, B. (1980) Abundance, biomass and contribution to energy flow by soil nematodes in terrestrial ecosystems. *Journal of Nematology* 11:86–196.

Sohlenius, B., Wasilewska, L. (1984) Influence of irrigation and fertilization on the nematode community in a Swedish pine forest soil. *Journal of Applied Ecology* 21: 327-342.

Sohlenius, B., Boström, S., Viketroft, M. (2011) Effects of plant species and plant diversity on soil nematodes – a field experiment on grassland run for seven years *Nematology* 13: 115-131.

Stamou, G.P., Papatheodorou, E. M., Hovardas, A., Argyropoulou, M. D. (2005) Some structural and functional characteristic of a soil nematode community from a Mediterranean grassland. *Belgian Journal of Zoology* 135: 253-259.

Stanton, N.L., Morrison, D., Laycock, W.A. (1984) The effect of phytophagous nematode grazing on blue grama die-off. *Journal of Range Management* 37: 447–450.

Sturhan, D., Lišková, M. (2004) Cyst nematodes in the Slovak Republic. *Helminthologia* 41: 271-219.

Šalamún, P., Renčo, M., Miklisová, D., Hanzelová, V. (2011) Nematode community structure in the vicinity of a metallurgical factory. *Environmental Monitoring and Assessment* (in press). DOI: 10.1007/s10661-011-1932-y.

Tomar, V.V.S., Ahmad, W. (2009) Food web diagnostics and functional diversity of soil inhabiting nematodes in a natural woodland. *Helminthologia* 46: 183-189.

Urzelai, A., Hernández, A.J., Pastor, J. (2000) Biotic indices based on soil nematode comunities for assessing sil duality in terrestrial ecosystems. *The Science of the Total Environment* 247: 253-261.

Ustinov, A.A., Zinovev, V.G. (1963) Važnejšie fitogeľminty USSR. In *Geľminty čeloveka, životnych i rastenij i borba s nimi.* Moskva, Izdaťeľstvo Akademii Nauk SSSR, pp. 503-504.

Valocká, B., Sabová, M. (1997a) Structure of soil nematode communities in some region sof Slovakia with different pollution level. *Biologia, Bratislava* 52: 609-613.

Valocká, B., Sabová, M. (1997b) Communities of soil and plant parasitic nematodes in two types of grass land. *Helminthologia* 34: 97-103.

Valocká, B. Sabová, M., Renčo, M. (2001) Soil and plant nematode communities of two types of ecosystems. *Helminthologia* 38: 105-109.

Verschoor, B.C., de Goede, R.G.M., de Vries, F.W. Brussaard, L. (2001) Changes in the composition of the plant-feeding nematode community in grasslands after cessation of fertiliser application. *Applied Soil Ecology* 17:1–17.

Viketoft, M., (2007) Plant induced spatial distribution of nematodes in a semi-natural grassland. *Nematology* 9: 131-142.

Viketoft, M., (2008) Effects of six grassland plant species on soil nematodes: A glasshouse experiment. *Soil Biology & Biochemistry* 40: 906–915.

Viketof, M., Bengtsson, J., Sohlenius, B., Berg, M.P., Petchey O. Palmborg, C., Huss-Danell, K. (2009) Long-term effects of plant diversity and composition on soil nematode communities in model grasslands. *Ecology* 90: 90–99.

Walter, D.E. (1988) Nematophagy by soil arthropods from the shortgrass steppe, Chihuahuan desert and Rocky Mountains of the central United States. *Agriculture, Ecosystems & Environment* 24: 307-316.

Wang, K.H. McSorley, R., Bohlen, P. Gathumbi S.M. (2006) Cattle grazing increases microbial biomass and alters soil nematode communities in subtropical pastures. *Soil Biology & Biochemistry* 38: 1956–1965.

Wardle, D.A. (2002) *Communities and ecosystems: linking the aboveground and belowground components.* Princeton University Press. Princeton, New Jersey. 392 p.

Wardle, D., Lavelle, P. (1997) Linkages between soil biota, plant litter quality, and decomposition. In: Cadish, G., Giller, K.E. (ed,), *Driven by Nature*. CAB International, Wallingford, pp 107-124.

Washington, H.G. (1984) Diversity, biotic and similarity indices: a review with special relevance to aquatic ecosystems. *Water Research* 18:653-649.

Wasilewska, L. (1971) Trophic classification of soil and plant nematodes. *Wiadomosci Ekologiczne* 17: 379-388.

Wasilewska, L. (1976) The role of nematodes in the ecosystem of meadow in Warsaw environs. *Polish Ecological Studies* 2: 137-156.

Wasilewska, L. (1994) The effect of age of meadows on succession and diversity in soil nematode communities. *Pedobiologia* 38: 1–11.

Wasilewska, L. (1995) Differences in development of soil nematode communities in singe and multi-species grass experimental treatments. *Applied Soil Ecology* 2: 53-64.

Wasilewska, L. (1997) Soil invertebrates as bioindicators, with special reference to soil-inhabiting nematodes. *Russian Journal of Nematology* 5: 113 – 126.

Yeates, G.W. (1979) Soil nematodes in terrestrial ecosystems. *Journal of Nematology* 11: 213-229.

Yeates, G.W. (1982) Variation in soil nematode diversity under pasture with soil and year. *Soil Biology and Biochemistry* 16: 95-102.

Yeates, G.W. (1994) Modification and qualification of the Nematode Maturity Index. *Pedobiologia* 38: 97-101.

Yeates, G.W. (1999) Effect of plants on nematode community structure. *Annual Review of Phatopathology* 37:127-149.

Yeates, G.W. (2003) Nematodes as soil indicators: functional and biodiversity aspects. *Bioliogy and Fertility of Soils* 37:199–210.

Yeates, G. W., Bongers, T., De Goede, R. G. M., Freckman, D. W., Georgieva, S. S. (1993) Feeding habits in soil nematode families and genera – an outline for soil ecologists. *Journal of Nematology* 25: 315–331.

Yeates G W, Bardgett R D, Cook R, Hobbs P J, Bowling P J and Potter J F (1997) Faunal and microbial diversity in three Welsh grassland soils under conventional and organic management regimes. *Journal of Applied Ecology* 34: 453–470.

Yeates, G.W., Orchard, V.A., Speir, T.W., Hunt, J.L., Hermans, M.C.C. (1994) Impact of pasture contamination by copper, chromium, arsenic timber preservative on soil biological activity. *Biology and Fertility of Soils* 18: 200-208.

Zaborski, E. (2009) *Soil nematodes, http://www.extension.org/article/24726.*

Zlotin, P.I. (1969) Roľ bezpozvonočnych životnych v mineralizacii rastiteľnogo opada. Materialy tretjego vsesojuznogo soveščanija. *Kazaň: Nauka*, 74-77

Zolda, P. (2002) Untersuchungen zur Nematodenfauna der Trockengrasen im Nationalpark Neusiedler See-Seewinkel. *Verhandlungen des Zoologisch-botanischen Vereins Gesellschaften Osterreichs.* 139: 53-58.

Zolda, P. (2006) Nematode communities of grazed and ungrazed semi-natural steppe grasslands in Eastern Austria. *Pedobiologia* 50: 11-22.

INDEX

C

D

E

F

G

H

I

J

O

P

Q

R

S

T

Y